Cell and Developmental Biology
of the Eye

Cell and Developmental Biology
of the Eye

Series Editors
Joel B. Sheffield and S. Robert Hilfer

Ocular Size and Shape: Regulation During Development

Cellular Communication During Ocular Development

The Proceedings of the Philadelphia Symposia on Ocular
and Visual Development

Cellular Communication During Ocular Development

Edited by
Joel B. Sheffield and
S. Robert Hilfer

With 64 Figures

Springer-Verlag
New York   Heidelberg   Berlin

Dr. Joel B. Sheffield
Dr. S. Robert Hilfer
Department of Biology
Temple University
Philadelphia, Pennsylvania 19122, U.S.A.

Sponsoring Editor: Philip Manor
Production: Marie Donovan

On the front cover: Induced neovascularization on the chick CAM. See page 149.

Library of Congress Cataloging in Publication Data
Main entry under title:
Cellular communication during ocular development.
  (Cell and developmental biology of the eye)
  Papers from the Sixth Symposium on Ocular and Visual Development,
held June 1981 in Philadelphia, Pa.
  Bibliography: p.
  Includes index.
  1. Eye—Congresses.   2. Cell interaction—Congresses.   I. Sheffield,
Joel B.   II. Hilfer, S. Robert.   III. Symposium on Ocular and Visual
Development (6th:1981:Philadelphia, Pa.)   IV. Series.   [DNLM:
1. Cell communication—Congresses.   2. Eye—Embryology—Congresses.
W3 SY5363 6th 1981c / WW 101 C394 1981| QL949.C44   1982
599'.01823   82–19251

ISBN-13: 978-1-4612-5766-0   e-ISBN-13: 978-1-4612-5764-6
DOI: 10.1007/978-1-4612-5764-6

Series Preface

The eye has fascinated scientists from the earliest days of biological investigation. The diversity of its parts and the precision of their interaction make it a favorite model system for a variety of developmental studies. The eye is a particularly valuable experimental system not only because its tissues provide examples of fundamental processes, but also because it is a prominent and easily accessible structure at very early embryonic ages.

In order to provide an open forum for investigators working on all aspects of ocular development, a series of symposia on ocular and visual development was initiated in 1973. A major objective of the symposia has been to foster communication between the basic research worker and the clinical community. It is our feeling that much can be learned on both sides from this interaction. The idea for an informal meeting allowing maximum exchange of ideas originated with Dr. Leon Canbeub, who supplied the necessary driving force that made the series a reality. Each symposium has concentrated on a different aspect of ocular development. Speakers have been selected to approach related topics from different perspectives.

This book series, "Cell and Developmental Biology of the Eye," marks the beginning of the publication of the Philadelphia symposia on ocular and visual development on a regular basis. A previous volume, *Ocular Size and Shape: Regulation During Development*, was published by Springer-Verlag in 1981. We hope that the introduction of this proceedings series will make the results of research on ocular cell and developmental biology more widely known and more easily accessible.

Preface

The Sixth Symposium on Ocular and Visual Development was held in June 1981. The major emphasis of the meeting was on cellular communication during development. Three areas of active research were discussed: synaptogenesis, cell coupling, and humoral factors. In the last few years, exciting observations have been made on the way synapses form within the retina, how optic nerve fibers find their way to the tectum, how cells communicate with each other to direct their specialization or to guide their pathways, and in the way soluble factors exert an effect on development over great distances. It has been a tradition of the Symposium on Ocular and Visual Development to include a presentation on a rapidly developing area of clinical practice. This year's paper concerned recent advances in improving the sight of individuals with low vision.

This volume arose from the papers that were presented at the meeting. We are indebted to the session organizers, Dr. E. Raviola, Dr. M.V.L. Bennett, and Dr. D. Beebe for their expertise in selecting the speakers and chairing the sessions, to the speakers for the stimulating presentations and their contributions to this volume, and to the reviewers of the manuscripts for their helpful comments.

The symposium could not have been held without the generous financial backing of the Temple University College of Liberal Arts and the Pennsylvania College of Optometry. We also thank Merck, Sharp and Dohme, Inc., for their donation. This volume was prepared with the expert guidance of Dr. Philip Manor and Marie Donovan of Springer-Verlag, and the skills of Jo-Ann Felder and Michael Czeredarczuk who helped prepare the text and many of the figures. We thank them for their help.

June 22, 1982   Joel B. Sheffield
Philadelphia, Pennsylvania   S.R. Hilfer

Contents

Contributors

S. Appel, William Feinbloom Vision Rehabilitation Center, The Eye Institute, Pennsylvania College of Optometry, Philadelphia, PA 19141, U.S.A.

R. Baughman, Department of Neurobiology, Harvard Medical School, Boston, MA 02115, U.S.A.

M.V.L. Bennett, Department of Neuroscience, Rose F. Kennedy Center, Bronx, NY 10461, U.S.A. *and* Marine Biological Laboratory, Woods Hole, MA 02543, U.S.A.

R. Brilliant, William Feinbloom Vision Rehabilitation Center, The Eye Institute, Pennsylvania College of Optometry, Philadelphia, PA 19141, U.S.A.

R.F. Dacheux, Department of Anatomy, Harvard Medical School, Boston, MA 02115, U.S.A.

L.J. Fisher, Department of Ophthalmology, Henry Ford Hospital, Detroit, MI 48202, U.S.A.

C.S. Goodman, Department of Biological Sciences, Stanford University, Stanford, CA 94305, U.S.A.

A.L. Harris, Department of Neuroscience, Rose F. Kennedy Center, Bronx, NY 10461, U.S.A. *and* Marine Biological Laboratory, Woods Hole, MA 02543, U.S.A.

P.T. Janssen, Laboratorium voor anatomie en embryologie, Rijksuniversiteit Utrecht, Utrecht BK 3512, The Netherlands

R.J. Mello, Wilmer Ophthalmological Institute, Johns Hopkins University, School of Medicine, Baltimore, MD 21205, U.S.A.

J.A. Raper, Department of Biological Sciences, Stanford University, Stanford, CA 94305, U.S.A.

H. Rothstein, Department of Biological Sciences, Fordham University, Bronx, NY 10458, U.S.A.

S.C. Sharma, Department of Ophthalmology, New York Medical College, Valhalla, NY 19595, U.S.A.

D.C. Spray, Department of Neuroscience, Rose F. Kennedy Center, Bronx, NY 10461, U.S.A. and Marine Biological Laboratory, Woods Hole, MA 02543, U.S.A.

H. van der Starre, Laboratorium voor anatomie en embryologie, Rijksuniversiteit Utrecht, Utrecht BK 3512, The Netherlands

M. van der Starre-van Bekkum, Laboratorium voor anatomie en embryologie, Rijksuniversiteit Utrecht, Utrecht BK 3512, The Netherlands

W.J. van Doorenmaalen, Laboratorium voor anatomie en embryologie, Rijksuniversiteit Utrecht, Utrecht BK 3512, The Netherlands

A. Weinsieder, Department of Ophthalmology, University of Louisville, Louisville, KY 40292, U.S.A.

B. Worgul, Department of Ophthalmology, Columbia University, College of Physicians and Surgeons, New York, NY 10032, U.S.A.

The Development of Specificity of Retinal Central Connections: Changing Concepts

S.C. Sharma

> From the functional point of view the growth cone
> of the retinal ganglion cells may be regarded as a sort
> of club or battering ram, endowed with exquisite
> chemical sensitivity, with rapid amoeboid movements,
> and with a certain impulsive force, thanks to which it
> is able to press forward and overcome obstacles met in
> its way, forcing cellular interstices until it arrives
> at its destination.
>
> Ramon y Cajal, 1899, pp. 544-545

The above-cited statement describes perhaps most elegantly
the initial concept of how neurons may form synaptic connections.
In the decades following Cajal's observation, it was recognized
that vertebrate neurons send their axons in the embryonic CNS to
form highly specific point-to-point connections with specific
types of target neurons at particular loci in the brain (for
review, see Gaze, 1970; Jacobson, 1978; and Lund, 1978). For the
present review, fundamental work apparently started with Sperry's
observation of the early 1940's. Sperry (1943, 1944) found that
when the optic nerve of lower vertebrates was cut or crushed, the
optic fibers regenerated back to their original position. This
conclusion was based upon behavioral testing of the animals
following eye rotation and consideration of other surgical
procedures. Later Sperry (1950) wrote "the optic fibers differ
from one another in quality according to the particular locus of
the retina in which the ganglion cells are located. The retina
apparently undergoes a polarized field-like differentiation within
the retina and the tectum during development, which brings about
specification of the ganglion cells and the tectal cells." The
functional relations established by the optic axon in the brain
centers are patterned in a systematic manner.

In essence Sperry (1945) postulated that specific regrowth of axons following optic nerve section was mediated by specific chemoaffinities. Implicit in this particular notion was an idea that during development, cells of the retina and tectum acquire cytochemical labels denoting their position within the retina or the tectum. Growing axons then read these labels by a kind of surface chemotaxis and are thus guided to the tectal cells having corresponding labels. Subsequent experiments by Sperry and co-workers on the retinotectal system apparently gave credence to his theory and indicated that the matching process was very precise and rigid. However, most of the support for Sperry's theory came from indirect evidence, primarily behavioral studies of animals whose eye cup had been rotated 180° at various stages of development. Eyes so manipulated developed normal vision if the operations were performed at a sufficiently early stage; whereas inverted visuomotor behavior developed if eyes were rotated at later stages.

The studies of Szekely (1954) suggested that the nasotemporal axis of the eye was fixed before the dorsoventral axis. These studies were further tested electrophysiologically by Jacobson (1968a) who identified a critical period in the developing eye of the embryo organizing the future visual pathways of the animal. In Xenopus this critical period was between stage 28-31. When the optic cup was rotated through 180° at stage 28/29, the toad developed a normal visual projection. When similar operations were performed at stage 30, the visual projection was abnormal in the sense that the map in the mediolateral direction across the tectum was normal (i.e. dorsoventral across the retina) but inverted in the rostrocaudal direction across the tectum (anteroposteriorly across the retina). Eyes rotated at stage 31 or later showed inverted maps on both directions across the tectum (Jacobson, 1968a). Jacobson (1968b) also demonstrated that the first ganglion cells in the retina become post-mitotic shortly before the specification of the ganglion cells' central connections. Hence, the cessation of DNA synthesis in the differentiating ganglion cells was followed closely by the specification of the cells' central connections.

In a detailed series of subsequent experiments, Jacobson, in collaboration with Dr. R.K. Hunt, suggested that in the long sequence of "developmental programming" events, the critical period is but one irreversible step. In one experiment these

authors showed that when the presumptive optic vesicle reaches the
periphery of the stage 22/23 Xenopus embryo, it already possesses
a pair of primitive retinal axes, basically aligned with
anteroposterior and dorsoventral axes of the eye. These axes are
stable (Hunt and Jacobson, 1973a). However, until the eye reaches
stage 28, the primitive retinal axes are replaceable. In 2 to 6
hours following 180° eye rotation, the existing axes are replaced
by a new pair or orthogonal axes which are themselves "stable but
replaceable" and which are once again aligned with the major axes
of the embryo's body (Hunt & Jacobson, 1972a, 1974).

In just 2 steps as the optic cup develops from stage 28 to
31, the capacity to undergo axial replacement is lost and the
existing "replaceable axes" are locked-in and specified as
permanent reference axes for the generation of locus specificity
in the retina (Hunt and Jacobson, 1972a, b). These authors
further concluded that after stage 31, the specified axes and
subsequently their ganglion cell central connections cannot be
modified by either a) rotation of the eye transplanted into a
pre-stage 28 host orbit; b) prolonged organ culture of the eye;
c) serial reintroduction into younger embryos; d) deprivation of
the retina from its central connections for 4-6 weeks; or
e) disruption of the normal pattern of retinal growth (Hunt and
Jacobson, 1972a, b; Hunt, 1975).

The relationship between the post-mitotic state of retinal
ganglion cells and the time when the ganglion cell central
connections become specified has been shown in only one species
(Xenopus, Jacobson, 1968). Since these remarkable observations
are of central importance in understanding the principles of
specification of neuronal connections, Dr. Hollyfield and I were
prompted to test whether the earlier observations of Xenopus are
valid in a different species, Rana pipiens. We studied the
cessation of DNA synthesis in the retinal rudiment in order to
determine the time the first ganglion cells become post-mitotic.
We also examined the retinotectal projection maps from eyes
rotated at various stages of development in order to determine the
time of specification of the retinal axes.

In direct contrast to the studies of Hunt and Jacobson, our
experiments exploring the timing of axial specification in the
developing retina of Rana pipiens showed that ganglion cell
central connections were specified before the first ganglion cells
became post-mitotic (Sharma & Hollyfield, 1974a). We observed

that specification of the retina had occurred prior to early
tail-bud stages while all cells in the retinal neuroepithelium
continue to divide. Furthermore, the visual maps in these studies
with Rana could be predicted prior to actual electrophysiological
mapping by the position of the choroidal fissure on the rotated
eye. In the normal animal a small notch or cleft is present along
the ventral margin of the iris. Closely associated with the
fissure position is a large blood vessel on the inferior aspect of
the globe. When these morphological markers were present in any
position out of register with their normal alignment, the
retinotectal map was always shifted through the same angle as was
the eye (Sharma & Hollyfield, 1974a). Although this prominent
ventral marker is also present in Xenopus laevis, none of the
earlier studies by Jacobson (1968a, b) or Hunt & Jacobson (1972a,
b; 1973a, b) has commented on the relative position of the
choroidal fissure in relation to orientation of the retinotectal
map.

The consistency of our results in Rana pipiens prompted us to
reinvestigate in Xenopus the orientation of retinotectal maps
relative to choroidal fissure position in eyes rotated at various
stages of development or reciprocally exchanged between left and
right orbit (Sharma & Hollyfield, 1974b, 1980). The position of
choroidal fissues in Xenopus laevis is prominent at tadpole stages
as well as in postmetamorphic animals when all of our
electrophysiological mapping was performed. The choroidal fissure
was prominent in each of the experimental animals used in these
studies.

The normal visual field of each eye in Xenopus projects to
the contralateral optic tectum; the nasal visual field to the
rostral tectum; temporal field to the caudal tectum; dorsal field
to the medial tectum; and inferior field to the ventral tectum
(See Fig. 1a). In Xenopus whose eye was rotated 180° between
embryonic stage 24-28, the resultant visuotopic projection was
also rotated 180° (Fig. 1b). This result was consistent
regardless of the stage of eye rotation. However, in very few
cases, duplicate maps were found. In these maps, the dorsal
visual hemifield projected to the extent of the dorsal tectum;
whereas, the ventral hemi-visual field projected to the extent of
the dorsal tectum but was rotated by 180°. These animals had two
choroidal fissures, one directed dorsally and the other ventrally.

In animals where eyes were reciprocally exchanged with 0°

rotation, the choroidal fissure appeared at its normal ventral
position. In these cases, the nasotemporal axis of the eye was
reversed and the dorsoventral axis was normal. The resultant map
was rotated 180° in the nasotemporal axis but was normal in the
dorsoventral axis (Fig. 1c). In animals with 180° reciprocal eye
exchange, the nasotemporal axis of the maps was normal, but the
dorsoventral axis was rotated (Fig. 1d). In both instances the
projections were consistent with the position of the eye, whether
the exchanges were done at stage 27 or later. Similar results
have been obtained by Gaze and his co-workers (Gaze et al., 1979)
in Xenopus.

From these studies, the following conclusions can be made:
a) the degree of rotation of the visuotectal map always
corresponds to the degree of rotation of the choroidal fissure;
b) the reciprocal eye exchange, with or without eye rotation,
expresses retinotectal connections according to the original axes
of the eye and were never respecified by axial cues from the new
orbit; c) eyes reciprocally exchanged to contralateral orbit
projected invariably to the contralateral optic tectum indicating
the absence of side specificity.

Recently published studies by Gaze et al., (1979) have also
found that the results of visuotectal map orientation always
corresponded to the orientation of the eye at the time of mapping
except for a few cases in which "compound" maps were recorded. In
the latter cases, the portion of the map which was oriented like
the eye came from the originally operative eye tissue, while the
other portion of the map came from eye tissue apparently newly
grown from the optic stalk.

Using chimeras (albino/wild type) of Xenopus, Gaze et al.,
(1979) showed that resultant eyes were of "compound type" with two
fissures. The original fissure of the dorsal eye was graft
derived, whereas the ventral eye with a normal fissure was derived
from the host tissue.   Gaze et al., suggested that the normal
ventral eye of the "compound eye" was probably derived from the
optic stalk of the host, giving rise to a normal map.

The evidence of ocular respecification following eye rotation
in the chick embryo (Goldberg, 1976a) could be the consequence of
a new fissure and the development of ventral retina from the optic
stalk.

The question of the position of the choroidal fissure is an
important aspect of recent studies. From the time of its

appearance in the ventral rim of the early optic cup, its location is evident through the life of the animal. It is not known when choroidal fissure position is determined in Xenopus laevis. In histological preparation of Xenopus retinas, the fissure can be identified as early as stage 27, but it is not readily apparent in the living embryo until the appearance of melanin in the pigment epithelium of stages 31 and 32. The determination of choroidal fissure position has been discussed elsewhere (Sharma & Hollyfield, 1974a).

The recent elegant studies of Holt (1980) showed that during normal development of the eye in Xenopus, the optic stalk cells migrate outward and upward during the optic cup stage of the eye and that these cells form the ventral retina where the future choroidal fissure appears. In those experiments, the eye rudiment at stages 22-37 was excised either partially or completely and was incubated in a medium containing ^{3}H-thymidine. After washing, the eye anlage was replaced in the socket of the same embryo. These embryos were fixed at stages 36-42. When the ventral half eye anlage at stage 24-27 was treated, the labelled cells appeared in the upper half of the ventral retina. Cells surrounding the choroidal fissure were unlabelled. In a second series of experiments, when the whole eye (stage 22-28) was incubated, the resultant autoradiographically labelled cells were present all

Fig. 1a Diagrammatic representation of the visuotopic map from the normal right eye to the left optic tectum. The abbreviations for this and for subsequent figures are the same: N, nasal; T, temporal; S, superior; I, inferior; R, rostral; C, caudal; M, medial; and L, lateral. In the normal map, the nasal visual field projects to the rostral tectum and the temporal field to the caudal tectum (represented by a heavy line with open arrow). The superior visual field is represented onto the medial tectum and the central and inferior visual field onto the lateral and ventral tectum (represented by thin line with closed arrow).

Fig. 1b Right visual field map onto the left tectum in an animal whose eye was rotated 180° at embryonic stage 24. This map is rotated 180°.

Fig. 1c The diagrammatic representation of the visuotopic map in which a left eye was transplanted into the right orbit at stage 27 without rotation. In this case, N-T axis of the field was rotated but the dorsoventral axis was normal (compare with Fig. 1-a).

Fig. 1d In this case the left eye was transplanted into the right orbit at stage 27 with 180° rotation. The dorsoventral axis was rotated but the N-T axis was normal.

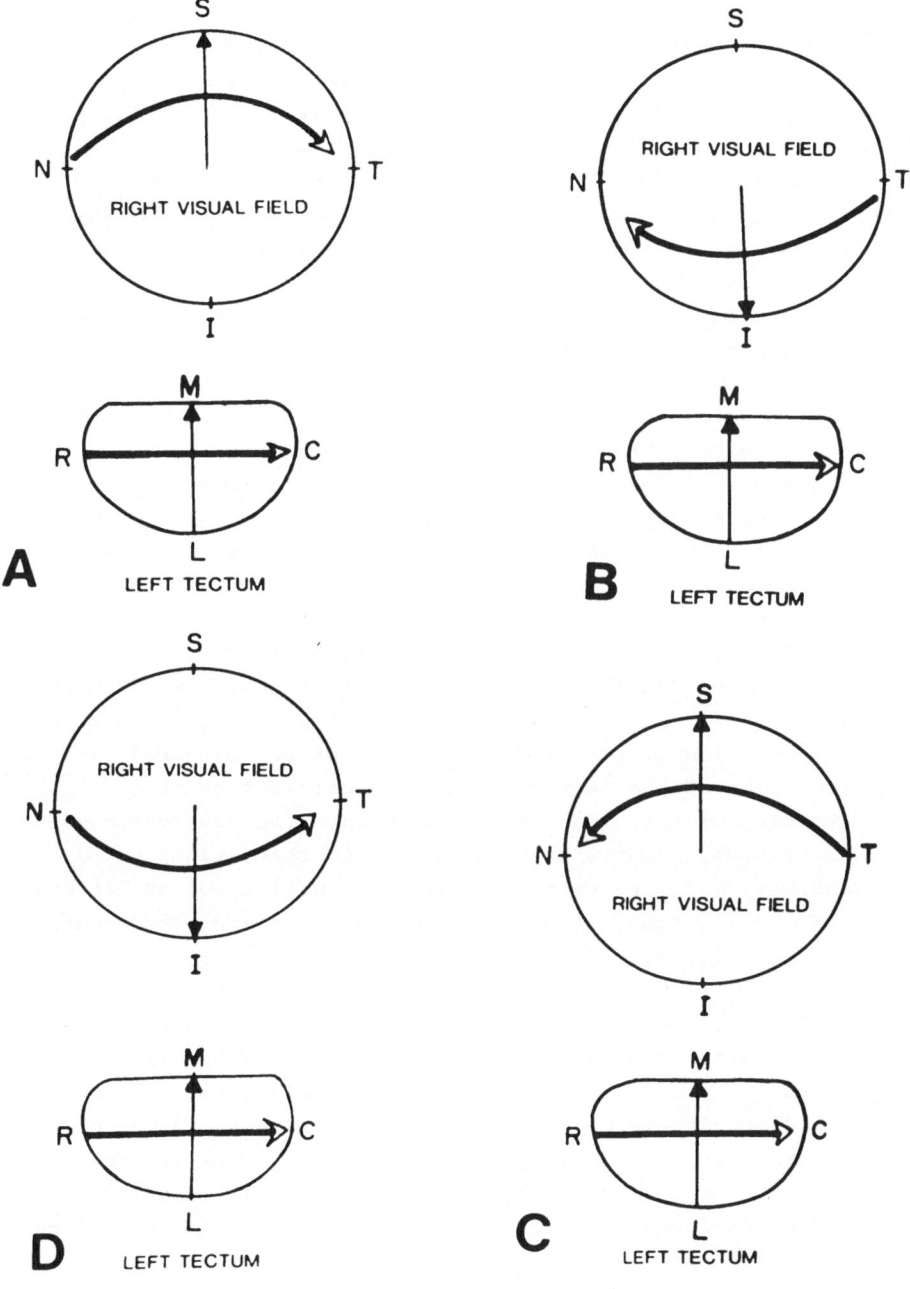

over the eye except in the ventral most retina containing the choroidal fissure. Unlabelled cells in the ventral retina were not found in studies where the eye was incubated after fissure formation was completed. This study showed, therefore, that there is a progressive movement of cells from the ventral towards the dorsal aspect of the retina during the optic cup formation stage.

From these results, it was inferred that in 180° eye rotation experiments, different fissure positions in the eye could be attributed to the extent of the surgery and the stage at which such operations were performed. The dorsally positioned fissure could be a consequence of deeper surgery at the optic vesicle stage or shallow surgery at the early eye cup stage. If, however, surgery was shallow at the optic vesicle stage, a normal ventral choroidal fissure would develop, assuming, of course, that stalk cells giving rise to ventral retina were still intact and that healing was excellent. When, during surgery, the stalk cell population was divided, a double fissure would result. It was further inferred that incomplete eye rotation leading to the formation of double fissures (a stage which has been considered by previous workers as the stage where the polarity of the eye is thought to be determined -- Sato, 1933; Beckwith, 1927; Stone, 1966; Goldberg, 1976a, b) may be the explanation for calling the early eye stages "labile" for the determination of axes of the eye.

The question of timing of cessation of DNA synthesis in the ganglion cells of the central retina in relation to the specification of their central connections also deserves comment. Jacobson (1968b) reports that all cells in the retinal neuroepithelium incorporate ^{3}H thymidine until stage 28, at which time a few cells drop out of the mitotic cycle and subsequently differentiate (become specified). Cima & Grant (1978) have recently reported that a bundle of axons, which they interpret as of ganglion cell origin, is present in the optic nerve head of Xenopus as early as stage 28. Since it is usual for axonal development to occur only in post-mitotic cells, it is unlikely that axons would be elaborated prior to withdrawal of these cells from the mitotic cycle. These observations of Cima and Grant suggest that some cells stop dividing much earlier than the stage reported by Jacobson (1968b). Furthermore, the polarization of the developing retina also occurs much earlier than previously reported by Jacobson (1968a). The exact relationship with respect to the relative timing of these two events is not clear at present.

Experimental evidence suggests that the retinotectal projections from eyes either rotated in the orbit or exchanged between left and right orbit were expressed according to the original axis of the eye at the time of manipulations. In no case was a normally oriented map obtained from a rotated eye when the choroidal fissure position indicated that the eye was indeed rotated. There is no compelling evidence to indicate that orbital cues can influence or respecify the axial coordinates within the retinal rudiment. There is also no evidence from recent studies that anterior-posterior and dorsal-ventral axes are specified as separate developmental events.

The preceding discussion leads to an obvious two-fold question: 1) how are retinal axes generated and 2) what role do the axes play in the final projection of the map over the extent of the tectum?

The answer to the second part of the question is that in every study conducted to date in the retinotectal system, the indication is that whatever one does with the tectum (remove, rotate, implant, explant, or partially ablate--Sharma, 1972, 1981; Schmidt, 1978), the resultant retinal map is always polarized; that is, the nasotemporal and dorsoventral axes of the projection are always maintained. This is perhaps essential for generation of visuotopic order in the projection.

The answer to the first part of the question is, on the other hand, somewhat more intricate, and I shall describe the few models and some experimental work which is pertinent here.

It has been argued that since the eye grows concentrically, i.e., radially, the spatially ordered sequence of cell division may specify the position of new cells. Hence, the specification of retinal cells according to a Cartesian coordinate system (orthogonal axes) can also be thought of in terms of a polar coordinate system (MacDonald, 1977). The position of a ganglion cell is defined by the distance from the center (a radial coordinate) and its angular distance or position along the circumference (the circumferential coordinate). The positional information of ganglion cells, accordingly, arises gradually during retinal development when cells become post-mitotic, thereby determining position by the radial distance from the center and angular coordinate by the angle of the fascicle that guides its axon to the optic nerve head. Each fascicle contains axons from an area in a common wedge-shaped sector of the retina, with oldest axons near the optic nerve head and youngest on the periphery. Therefore, it would be the first fascicle in the developing retina which would assign the angular coordinates for all the axons which will develop later. The spatio-temporal model for the development of optic nerve fiber topography is, therefore, consistent with the polar coordinate positional information hypothesis.

Bodick and Levinthal (1980) showed that a large optic bundle in developing zebra fish arises from cells in a pie-shaped sector of the retina and that cells from the opposite side of the choroidal fissure do not merge into the same bundle. The separation, leading to the formation of a crescent pattern by fibers leaving the eye, is a consequence of the barrier presented by the choroidal fissure. These authors suggested that the opening in the crescent pattern in the developing retina is important since it leads to inversion of the overall topography of the optic axons in the optic nerve, where youngest fibers are on one side of the nerve and the oldest fibers are on the opposite side. (See also Scholes, 1979). Bodick and Levinthal (1980) suggested that the polar coordinates of the cell body position in the retina correspond to the rectangular coordinates in the optic nerve because of the inversion of overall topography of optic axons in the nerve. They proposed that the ganglion cell axons follow pathways that are determined by the substrate along which they grow and that locus specificity of ganglion cells is determined by the time and place at which they initiate the growth of an axon. These rules, then, may perhaps give rise to (a) locus specificity of the ganglion cells and (b) the ordering of optic axons within the optic nerve. However, whether similar rules apply to the mode of selective synapse formation on the target centers remains to be determined.

Dr. Gunther Rager (see review, 1980) has proposed that the position of a ganglion cell can be specified by spatiotemporal coordinates with respect to the developing optic fiber layer. The ganglion cells formed in incremental rings of growth are transformed into a crescent at the entrance of the optic stalk during eye development. It was asserted that the mode of growth of the optic axons and the tectum were sufficient to the development of a retinotectal map. Rager's hypothesis for the connection formation includes a) temporal and spatial order;

b) gradient of cell generation and maturation; c) contact guidance; and d) cell death. This hypothesis does not require Sperry's initial suggestion of a chemoaffinity hypothesis.

In systems where aggregation and sorting of dissociated cells in the retina have been studied (for review, see Moscona, 1976), it has been shown that selective cell adhesion exists. For example, in the retinotectal system, Halfter, et al. (1981) showed that a gradient of adhesion between retinal and tectal cells exists. A preferential adhesion of tectal cell membranes in 6-day old chick embryonic retinal explants in cultures was shown to the nasal retinal ganglion cell axons but not from the temporal retina. Furthermore, the preferential adhesion of the retinal neurite was independent of the dorsoventral quadrant of the retina. These authors argued that the preferential adhesion by the nasal neurite suggested a continuous gradient in the nasotemporal retina. The molecular nature of the adhesion gradient, however, remained elusive.

A recent paper (Trisler et al., 1981) has, for the first time, demonstrated the existence of the concentration gradient of a particular molecule in the chick retina. These authors showed that a monoclonal antibody bound more strongly to the dorsal retina as compared to the ventral one. The dorsal retina binds 35-fold more to this antibody than does the ventral retina. These authors referred to the specific antigen which is detected by the monoclonal antibody as "Toponymic," i.e. "marker of the position." This probable cell-surface protein is present in all layers of the 14-day old chick retina. The binding gradient represented differences in the amount of antigen per cell. A similar gradient was observed as early as in a 4-day old embryonic eye. Furthermore, in one animal with an ectopic eye (situated in the middle of the forehead) with abnormal orientation, the toponymic molecule gradient on it had normal orientation with respect to its choroidal fissure. The last example indirectly supports our earlier suggestion that topography of the retina for its subsequent map is always aligned by the position of the eye with respect to its choroidal fissure.

The question of how the molecular gradient of the retina expresses itself in the optic tectum for the generation of a topographic map remains, at present, unclear.

In conclusion, recent experiments on eye rudiments give absolutely no reason to suppose that specificiation of the eye for subsequent map formation involves polarization of the eye rudiment in Cartesian coordinates. Rather, specification of the eye may be controlled by polar coordinates or some other spatio-temporal order of cell generation. I have here outlined current concepts of eye specification. It is not clear exactly how position-dependent properties for each ganglion cell are acquired both in the polar coordinate system as well as the formation of the topographic molecule gradient which is ordered in relation to the axis of the retina. Certainly, the monoclonal antibodies-antigen approach opens an exciting chapter in the determination of specification of retinotectal connections.

Supported by N.I.H grants EY 01426 and 1K04-EY 00101

REFERENCES

Beckwith, C.J. 1927. The effect of the extirpation of the lens rudiment on the development of the eye in Amblystoma punctatum with special reference to choroidal fissure. J. Exp. Zool. 49: 217-259.
Bodick, N. and C. Levinthal. 1980. Growing optic nerve fibers follow neighbours during embryogenesis. Proc. Natl. Acad. Sci., U.S.A. 77: 4374-4378.
Cima, C. and P. Grant. 1978. Ultrastructure evidence of early retinal ganglion cell differentiation in Xenopus laevis. Soc. for Neuroscience Abst. 4: 622.
Gaze, R.M. 1970. The Formation of Nerve Connections. Academic Press, New York.
Gaze, R.M., J.D. Feldman, J. Cook and S.H. Chung. 1979. The orientation of the visuotectal map in Xenopus: developmental aspects. J. Embryol. Exp. Morph. 53: 39-66.
Goldberg, S. 1976a. Progressive fixation of morphological polarity in the developing retina. Develop. Biol. 53: 126-127.
Goldberg, S. 1976b. Polarization of the avian retina. Ocular transplantation studies. J. Comp. Neurol. 168: 379-392.
Halfter, W., M. Claviez and U. Schwarz. 1981. Preferential adhesion of tectal membranes to anterior embryonic chick retina neurites. Nature. 292: 67-70.
Holt, C. 1980. Cell movement in Xenopus eye development. Nature. 287: 850-852.
Hunt, R.K. 1975. Developmental programming for retinotectal patterns. pp. 131-150. In "Cell Patterning", Ciba Foundation Symposium 29,, Elsevier, New York.
Hunt, R.K., and M. Jacobson. 1972a. Development and stability of positional information in Xenopus retinal ganglion cells. Proc. Natl. Acad. Sci. U.S.A. 69: 780-783.
Hunt, R.K. and M. Jacobson. 1972b. Specification of positional information in retinal ganglion cells of Xenopus: stability of the unspecified state. Proc. Natl. Acad. Sci. U.S.A. 69: 2860-2864.
Hunt, R.K. and M. Jacobson. 1973a. Specification of positional information in retinal ganglion cells of Xenopus: Assay system for analysis of the unspecified state. Proc. Natl.

Acad. Sci. U.S.A. 70: 507-511.
Hunt, R.K. and M. Jacobson. 1973b. Neuronal locus specificity: Altered pattern of spatial deployment in fused fragments of embryonic Xenopus eyes. Science. 180: 509-511.
Hunt, R.K. and M. Jacobson. 1974. Specification of positional information in retinal ganglion cells of Xenopus laevis: intra-ocular control of the time of specification. Proc. Natl. Acad. Sci. U.S.A. 71: 3616-3620.
Jacobson, M. 1968a. Development of neuronal specificity in retinal ganglion cells of Xenopus. Develop. Biol. 17: 202-218.
Jacobson, M. 1968b. Cessation of DNA synthesis in retinal ganglion cells correlated with the time of specification of their central connections. Develop. Biol. 17: 219-232.
Jacobson, M. 1978. Developmental Neurobiology. 2nd Ed. Plenum Press. New York.
Lund, R.D. 1978. Development and Plasticity of the Brain. Oxford University Press. New York.
McDonald, N. 1977. A polar coordinate system for positional information of the vertebrate neural retina. J. Theor. Biol. 69: 153-165.
Moscona, A.A. 1976. In "Neuronal Recognition" (ed. Barondes, J.S.) pp. 205ff Plenum Press, New York.
Rager, G.H. 1980. Development of the retinotectal projection in the chicken. Adv. Anat. Embryol. Cell. Biol. 63: 1-92.
Ramon y Cajal, S. 1899-1904. "Textura del sistema nervioso del hombre y del los vertebrados estudios sobre el plan estructural y composicion histologica de los centros nerviosos, adicimados de-consideraciones fisiologicas fundadas in los nuevos des cubrimientos." Vol. 1. 556 pp. Madrid. N. Moya.
Sato, T. 1933. Über die Determination der fetal Augenspalts bei Triton taeniatus. Arch. F. Entwmech. d. Org., Bd. 128: 342-377.
Schmidt, J.t. 1978. Retinal fibers alter tectal positional markers during the expansion of the half retinal projection in goldfish. J. Comp. Neurol. 177: 279-300.
Scholes, J.H. 1979. Nerve fiber topography in the retinal projection to the tectum. Nature 278: 620-624.
Sharma, S.C. 1972. Retinotectal connections of a heterotopic eye. Nature. 238: 286-287.
Sharma, S.C. 1981. Retinal projection in a non-visual area after bilateral tectal ablation in goldfish. Nature. 291: 66-67.
Sharma, S.C. and J.G. Hollyfield. 1974a. Specification of retinal central connections in Rana pipiens before the appearance of the first post-mitotic ganglion cells. J. Comp. Neurol. 155: 395-408.
Sharma, S.C. and J.G. Hollyfield. 1974b. The retinotectal projection in Xenopus laevis following right-left exchange of eye rudiment. Soc. for Neuroscience. Abst. 4: 421.
Sharma, S.C. and J.G. Hollyfield. 1980. Specification of retino-tectal connections during development of the toad Xenopus laevis. J. Embryol. Exp. Morph. 55: 77-92.
Sperry, R.W. 1943. Visuomotor coordination in the newt (Triturus viridescens) after regeneration of the optic nerves. J. Comp. Neurol. 79: 33-55.
Sperry, R.W. 1944. Optic nerve regeneration with return of vision in anurans. J. Neurophysiol. 7: 57-69.
Sperry, R.W. 1945. Restoration of vision after uncrossing of optic nerves and after contralateral transposition of the eye. J. Neurophysiol. 8: 15-28.

14

Sperry, R.W.   1950.   Neuronal specificity.   pp. 232-239.   In
    "Genetic Neurology" (P. Weiss, ed.). Univ. of Chicago Press,
    Chicago.
Stone, L.S.   1966.   Development, polarization and regeneration of
    the ventral iris cleft (remnant of choroid fissure) and
    protractor lentis muscle in urodele eyes.   J. Exp. Zool. 161:
    95-108.
Szekely, G.   1954.   Zür Ausbildung der lokalen funktionellen
    Spezifitat der Retina.   Acta Biol. Hung. 5: 157-167.
Trisler, G.D., M.D. Schneider and M. Nirenberg.   1981.   A
    topographic gradient of molecules in retina can be used to .
    identify neuron position.   Proc. Natl. Acad. Sci. U.S.A. 78:
    2148-2149.

Development of Retinal Synaptic Arrays in Mouse, Chicken, and Xenopus: A Comparative Study

Leslie J. Fisher

Amacrine, interplexiform and bipolar cells of the inner nuclear layer of the retina (Figure 1) form synapses which are arranged in a spatial configuration in the inner plexiform layer. I shall call this configuration an array. Qualitatively, among vertebrates, this array of synapses appears to be similar from species to species. Two major categories of synapses are always present: conventional synapses in processes of amacrine or interplexiform cells, and ribbon synapses in the processes of bipolar cells (Boycott et al., 1975; Dowling and Boycott, 1966; Dowling, 1968; Dubin, 1970; Raviola and Raviola, 1967). The unambiguous anatomical differences between these two categories has permitted a quantitative analysis of the IPL which goes beyond counting. Quantitative analysis has demonstrated that both absolute numbers of synapses and ratios of ribbon to conventional synapses differ from species to species (Dowling and Boycott, 1966; Dowling, 1968; Dubin, 1970; Fisher, 1979a,b; Fisher and Easter, 1979; Fisher, 1976; Tucker and Hollyfield, 1977).

An array has properties by which it can be described. These are properties which reside in the system of synapses and their spatial configuration rather than in an individual element of the array. Such properties as the density of synapses per unit volume of neuropil, or the mean distance between nearest neighboring synapses, or the number of synapses contained in a portion of retina are properties of the array rather than of its elements. Numerical descriptors of the arrays can be derived by a quantitative analysis of these properties. Moreover, descriptors, when determined at different stages of the animals' development, provide a quantitative measure of growth of the neuropil.

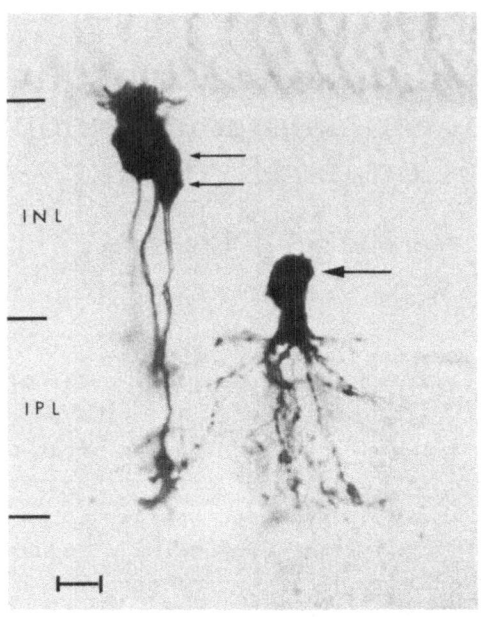

Figure 1. Golgi-stained amacrine cells (one arrow) and bipolar cells (two arrows) of the mouse retina. Calibration bar equals 10 µm. IPL = inner plexiform layer, INL = inner nuclear layer, GCL = ganglion cell layer. (Reprinted from Fisher, L.J., 1979.)

The studies reported here are an attempt to determine some of the numerical descriptors of the IPL arrays throughout the development of the retinas of mouse, chicken, and Xenopus. The way in which these descriptors change with time defines the life history of the IPL.

METHODS

The synapses formed by the pre-synaptic elements of the IPL can be identified by characteristic morphologic features. Two separate sub-arrays of synapses, those of ribbon synapses and those of conventional synapses can readily be discerned in electron micrographs. Ribbon synapses are associated with bipolar cells while conventional synapses are associated with amacrine and interplexiform cells.

The criterion used in this study to define the presence of a ribbon synapse was the identification of the electron-dense organelle with a surrounding halo of synaptic vesicles (Figure 2).

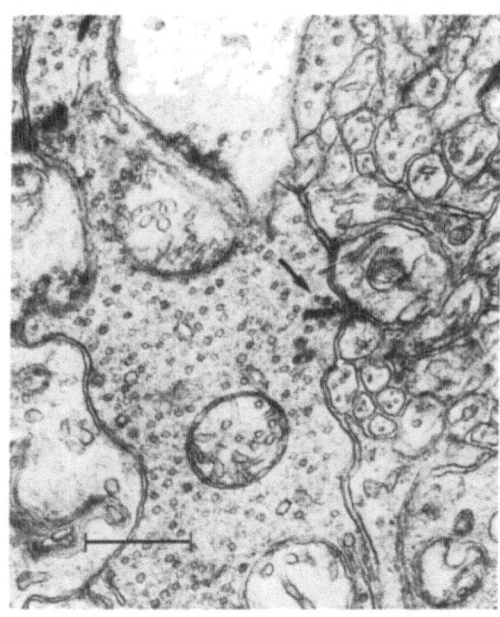

Figure 2. Electron micrograph of a ribbon synapse (arrow) in the
IPL of a mouse. Calibration bar equals 0.5 µm. (Reprinted from
Fisher, L.J., 1979.)

The organelle--called a ribbon--is actually disc-shaped and
surrounded by a halo of synaptic vesicles. When sectioned in
profile the ribbon usually, but not always, points between two
post-synaptic processes. Since the organelle can be cut in a
plane of section which does not include the plasma membrane,
ribbons can be counted accurately without requiring the positive
identification of post-synaptic membrane specializations.

The only processes of the IPL which contain ribbons are from
bipolar cells. In addition, the only type of synapse found in the
processes of bipolar cells are associated with ribbons. The array
of ribbon synapses formed by bipolar cells, therefore, can be
identified unambiguously (Dowling and Boycott, 1966; Dowling,
1968; Sidman, 1961).

Conventional synapses can be identified by the presence of
synaptic vesicles in close juxtaposition to a specialization of
pre-and-post-synaptic membranes (Figure 3). The criterion for
identification of conventional synapses in these studies was the
presence of a membrane specialization and at least 3 closely

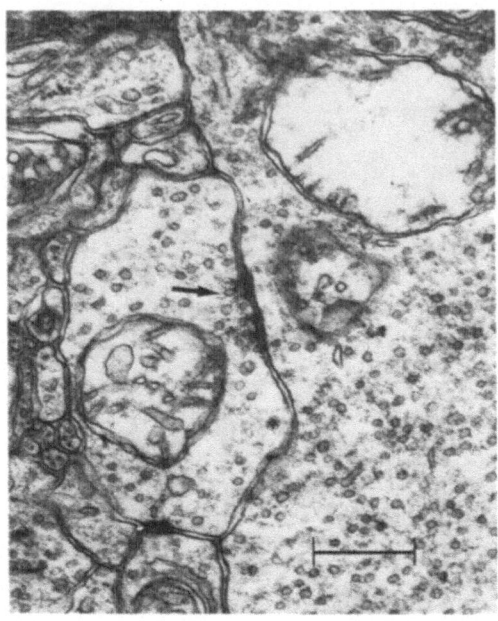

Figure 3. Electron micrograph of a conventional synapse (arrow) in the IPL of a mouse. Calibration bar equals 0.5 μm. (Reprinted from Fisher, L.J., 1979.)

subjacent vesicles. This is an arbitrary definition, but strict adherence to this as a working definition gave consistent results.

Just as ribbons are confined to the processes of bipolar cells, so too are conventional synapses of the IPL confined to the processes of amacrine or interplexiform cells. Since conventional synapses are the only type of chemical synapse found in these processes, the identification of these arrays is also unambiguous (Dowling and Boycott, 1966; Dowling, 1968; Raviola and Raviola, 1967).

Eyes were enucleated at 1400 +1 hour in all cases. After marking the eye for later orientation using 7-0 silk sutures, the eyes were hemisected, lentectomized and the vitreous body removed. The eye cups were fixed in a cacodylate-buffered aldehyde fixative, post-fixed in osmium, and embedded in epon.

The oldest part of the retina is the central portion adjacent to the optic nerve (Fujita, 1963; Hollyfield, 1971; Kahn, 1974; Sidman, 1961). Accordingly, the area of retina sampled in these studies was always within 500 microns of the optic nerve head.

Using the method of West (1972), the position within the retina of the tissue to be sampled could be precisely specified from animal to animal.

Electron micrographs of the entire thickness (width) of the IPL were prepared at a magnification of 26,000 times. On each of the micrographs all synapses were identified, marked, counted, and measured. The length of the ribbon organelle in ribbon synapses or the chord of the membrane specialization in conventional synapses were the characteristic lengths used in subsequent calculations. The area of IPL sampled was measured on the micrographs using a polar planimeter. The thickness of the tissue section was estimated from records of the interference color as noted during sectioning.

The Abercrombie correction as modified by Dubin (1970) was applied to all counts to yield the density of synapses per unit volume of IPL. The resulting statistic is the numerical density of synapses. This is simply the mean number of ribbon synapses or conventional synapses contained in 1 cubic micron of IPL, no matter what the total volume the IPL happens to be.

The IPL grows in thickness, and hence in volume, at the locus of sampling over the period covered by these studies. Since the numerical density of synapses can vary independently from the changes in volume of the neuropil, some other statistic is needed to determine the rate of formation of synapses and the total number of synapses at a sampling locus. The statistic used in these studies was the product of the IPL width and the numerical density of synapses. This product, expressed as synapses per unit area of retina, is termed the retinal planimetric density. It combines changes in the number of synapses per unit volume with the growth of the neuropil.

The physical significance of the planimetric density can be visualized by imagining a spot of light of unit area focused upon the IPL. The number of synapses in the IPL covered by this spot of light is equal to the planimetric density of synapses.

The chick embryos and the Xenopus tadpoles were staged by comparison with published criteria (Hamburger and Hamilton, 1951; Nieuwkoop and Faber, 1967). The resulting stages were used as the independent variable to define the points of measurement. Since temperature was controlled for both of these species during the period of growth sampled, the stages are proportional to age of the animal but in a non-linear manner.

Mice were not staged. Rather, they were sampled at specific ages from birth. The day of birth is taken as day zero of age. Pigmented, genetically wild type mice of the C57BL/6 strain were used.

OBSERVATIONS

Mouse

Growth curves of the retinal arrays were different for each of the species investigated. Figure 4 illustrates the numerical densities of both ribbon and conventional synapses of the mouse retina as a function of age after birth. Two conditions are immediately apparent: (1) the growth curves for the arrays of the two synaptic types display different time courses (but in both growth is followed by a plateau in numerical density), and (2) the initial appearance of the conventional synapses occurs before any ribbon synapses have formed.

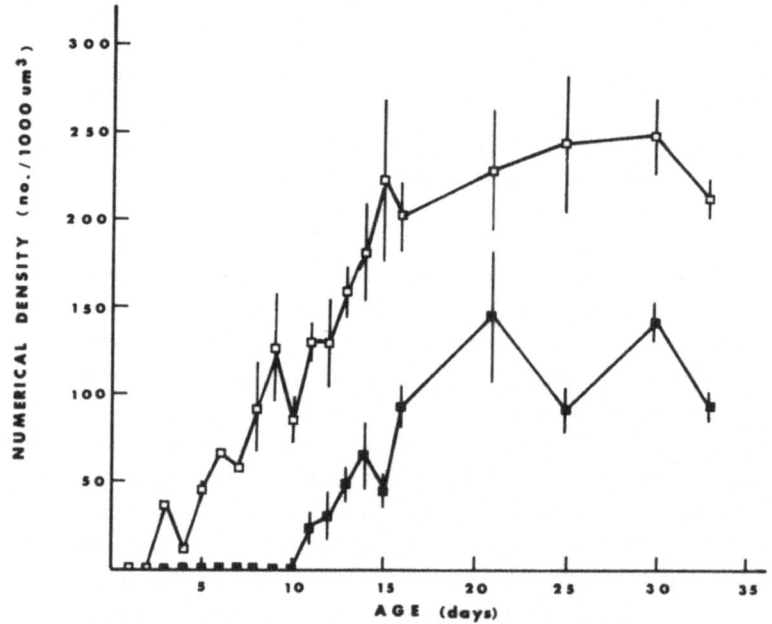

Figure 4. Numerical density of ribbon (filled symbols) and conventional (open symbols) synapses in the mouse inner plexiform layer as a function of post-natal age in days. The day of birth is counted as day zero. Numerical density is expressed as synapses per 1000 μm³. Error bars in this and all subsequent figures are ±1 SEM.

No chemical synapses of either the conventional or ribbon type are present before day 3 post-natal. On day 3, a few synapses that conform to the minimal criteria are present in the IPL. From day 3 to day 15 the numerical density of the array of conventional synapses increases. From day 15 to day 33 numerical density remains constant.

No ribbons are present until day 11 post-natal. Numerical density of the array of ribbons increases rapidly and at day 16 attains a value which is close to the mean of values measured thereafter.

The relative timing of the formation of the two arrays suggests a division in the growth of the mouse IPL into three phases: in the first only conventional synapses are formed, in the second phase both conventional and ribbon synapses are formed, and in the third phase the arrays of both ribbon and conventional synapses maintain a nearly constant numerical density.

Little change is seen in the shape of the growth curves of the arrays when they are expressed as planimetric densities (Figure 5). The number of synapses per unit area expresses the total number of synapses at that retinal locus, a figure which takes into account the growth of the IPL thickness. In these animals the IPL does not increase in thickness beyond day 18. The onset of phase three, therefore, coincides with the cessation of growth of the IPL. When the IPL attains its maximal thickness, there is no further increase in either numerical or planimetric densities of synapses.

Chicken

The retina of the chicken develops synaptic arrays before hatching. At stage 36 (day ten of incubation) the first conventional synapses are present. From stage 36 to stage 41 the numerical density of the array of conventional synapses increases (Figure 6). From stage 41 until hatching the numerical density remains nearly constant.

At stage 40 (day 14 of incubation) ribbons are first present. Unlike conventional synapses, the array of ribbon synapses continues to increase in numerical density until hatching.

The IPL of chick retina continually grows in thickness throughout embryonic life. When this growth of the neuropil is used to calculate the planimetric density of conventional synapses, the plateau seen in the numerical density disappears

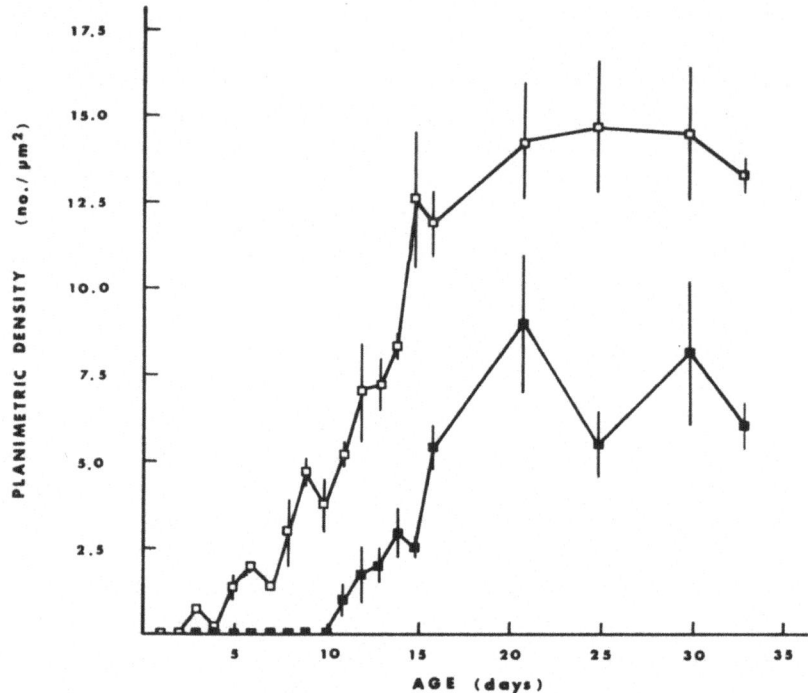

Figure 5. Planimetric density of synapses in the mouse IPL.
These curves are derived from the same data as Figure 1.
Planimetric density is equal to the number of synapses that would
be found in the IPL lying directly beneath a spot of light of unit
area focused on the retina. Planimetric density is expressed as
synapses per μm^2, age in days. Conventional synapses, open
symbols; ribbon synapses, filled symbols.

(Figure 7). The IPL adds conventional synapses to the array from
stage 36 until hatching. Ribbon synaptic arrays increase in
planimetric density from stage 40 to hatching.

The final values of numerical and planimetric densities
attained for both types of synapse are nearly equal in mouse and
chicken. In both species the plateau of numerical density for
conventional synapses is approximately 250 synapses per 1000 μm^3.
The value of numerical and planimetric densities of ribbons at
hatching for the chick and after day 15 for the mouse are also
equivalent: 150 ribbons per 1000 μm^3 and 7500 ribbons per 1000
per μm^2 for numerical and planimetric densities respectively.

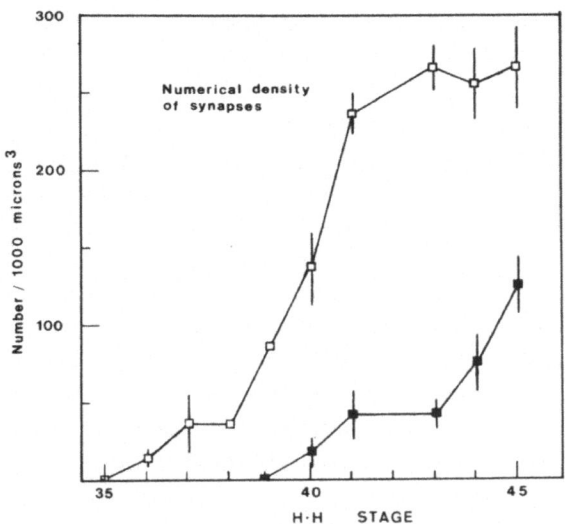

Figure 6. Numerical density of ribbon (filled symbols) and
conventional (open symbols) synapses in the IPL of the chick
retina as a function of the Hamburger-Hamilton stage of
development. Numerical density is expressed in synapses per
1000 μm^3.

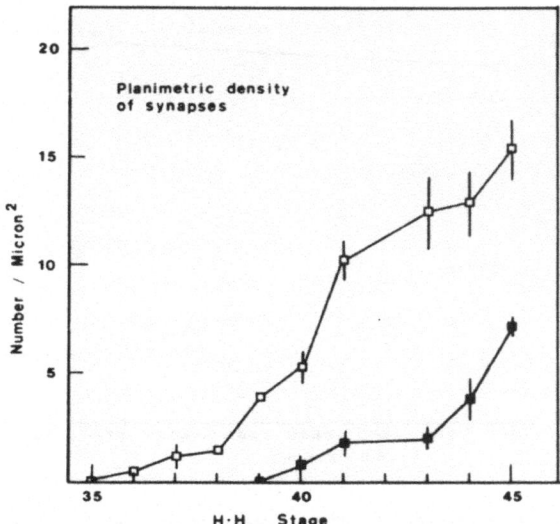

Figure 7. Planimetric density of ribbon (filled symbols) and
conventional (open symbols) synapses in the IPL of chick retina as
a function of Hamburger-Hamilton stage. Planimetric density is
expressed in synapses per μm^2.

The outer plexiform layer (OPL) and outer segments were also examined in this study. In the OPL, ribbons were first seen at stage 40. Outer segments were first present at stage 43.

Xenopus

Xenopus larvae rapidly develop a high numerical density of the arrays of both conventional and ribbon synapses (Figure 8). A sharp increase occurs from the nearly simultaneous appearance of ribbon and conventional synapses at stage 40 until stage 46, only 20 hours later. The values attained of 560 and 220 synapses per 1000 μm^3 for conventional and ribbon synapses respectively are nearly double the value measured in mouse and chicken.

Following the initial increase in numerical density the array of conventional synapses is maintained at a nearly constant value. The array of ribbon synapses decreases in density. Stage 57 marks metamorphosis. While the density of the array of conventional

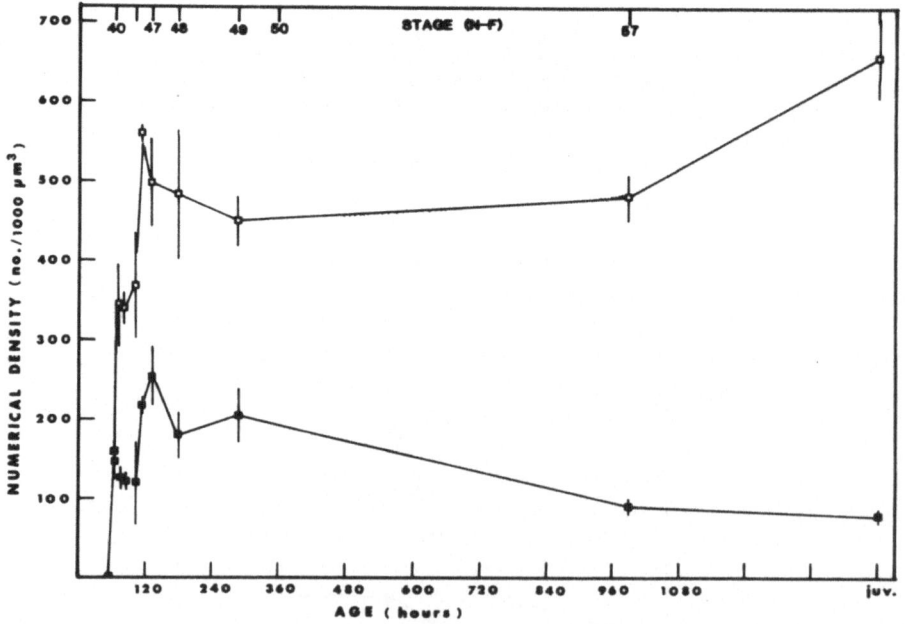

Figure 8. Numerical density of conventional and ribbon synapses in the IPL of Xenopus as a function of the Nieuwkoop and Faber stage. Numerical density is expressed as synapses/1000 μm^3. Conventional synapses, open symbols; ribbon synapses, filled symbols.

synapses is greater in the post-metamorphic animal, the density of the array of ribbon synapses is unchanged.

The thickness of the IPL grows throughout the larval life of Xenopus. Accordingly, the planimetric density of the array of conventional synapses increases markedly and the density of the array of ribbon synapses is more nearly constant (Figure 9). The planimetric densities indicate that conventional synapses are added to the array from stage 48 through hatching, and that ribbon synapses change only slightly in number. Since only one dimension of the IPL was measured in this study, the slight decrease in the planimetric density of the array of ribbon synapses might be due to an increase in volume of the IPL along the unmeasured dimensions.

Table I demonstrates the rapidity with which the synaptic arrays grow. For each unit of IPL the planimetric density of conventional synapses increases by adding one new synapse every 1-1/2 minutes in mice, every 1 minute in chick, and every 22

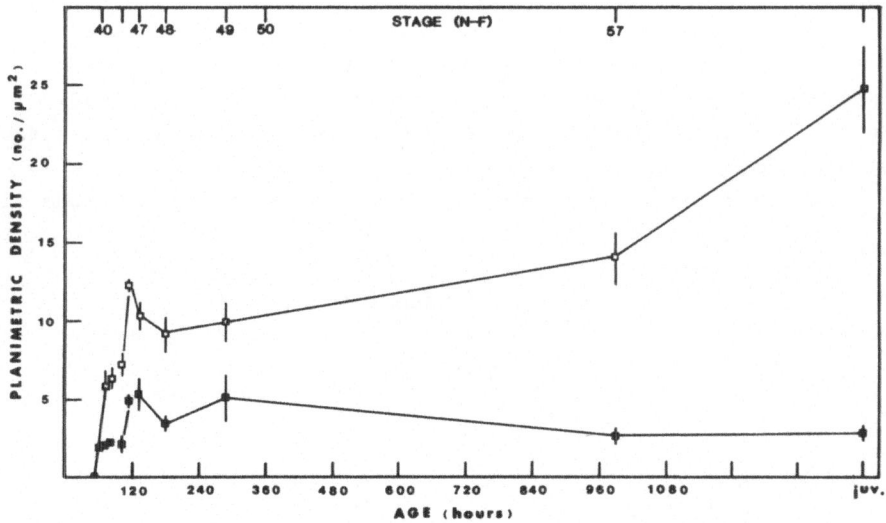

Figure 9. Planimetric density of conventional and ribbon synapses in the IPL of Xenopus as a function of the Nieuwkoop and Faber stage. Planimetric density is expressed as synapses per μm^2. Conventional synapses, open symbols; ribbon synapses, filled symbols.

seconds in Xenopus. Ribbons, too, are added at a rapid rate. In mouse, the planimetric density of ribbons increases by adding one new ribbon to every 1000 μm^2 of IPL each 1-3/4 minutes; chicken and Xenopus add ribbons even faster, at about one new synapse every minute.

DISCUSSION

Description of IPL arrays in terms of numerical and planimetric densities is a direct quantitative measure of two important features of a synaptic array: (1) the mean number of synapses in a unit volume and (2) the total number of synapses in a unit of IPL. When measured over time these descriptors define the growth of the synaptic array. An inter-specific comparison of the resultant growth curves suggests some unifying principles of growth and development of the retina.

Pioneering Synapses
In mouse and chicken the first synapses to appear are conventional synapses. The array of these synapses increases in density before any receptor to bipolar synapses are formed and before the outer segments of the receptors appear (Blanks, Adinolfi and Lolley, 1974; Fisher, 1979; McLaughlin, 1976; Meller and Tetzlaff, 1976; Meller and Tetzlaff, 1977; Olney, 1968; Olson, 1979). Thus, in both of these species these first synapses form with no normal sensory input.

Whether these pioneering synapses remain throughout the life

TABLE I

	Ribbon Synapses	Conventional Synapses
Mouse	105	90
Chicken	74	58
Xenopus	52	22

Table I. The mean time, in seconds, to form a synapse in 1000 m2 of retinal area during the period of maximal increase in planimetric density. The figures give the time taken to attain one new synapse in the IPL which would underlie a spot of light of 1000 μm^2 focused on the retina.

of the animal or are replaced by newer arrays as the IPL matures is not known. That they form before any sensory information can affect the array's configuration suggests that they have an organizational function in addition to any possible role in data processing. The primitive array formed by the pioneering synapses is modified throughout development by the continual addition of conventional synapses and the inclusion of ribbon synapses within the overall array.

A different sequence of appearance occurs in Xenopus. In Xenopus the arrays of both conventional and ribbon synapses develop concurrently. In this species, the initial development of synaptic arrays in the IPL occurs later than the initial development of the receptor outer segments. Synaptic organization of the outer plexiform layer matures at the same time as the arrays of the IPL. Thus in Xenopus there is a possibility of sensory input to the IPL during the formation of the first synapses. The initial organization of the Xenopus retina, however, occurs over an extremely short time interval. Detection of any pioneering synapses may have been prevented by the rapidity with which the neuropil grows.

First Ribbon Synapses

The time at which the first ribbons are present in the IPL of Xenopus, mouse, and chicken is also the time at which the first receptor to bipolar synapses are formed in the OPL (Chen and Witkovsky, 1978; Fisher, 1976; Hollyfield, 1971; Tucker and Hollyfield, 1977). The bipolar neurons, therefore, appear to make input and output synaptic connections simultaneously.

Numerical and Planimetric Densities

Comparison of the growth curves within the species shows that the two sub-arrays of synapses follow different time courses. In addition to this, the two arrays differ in the formation of new synapses during continued growth of the neuropil. The numerical densities and planimetric densities of each of the sub-arrays vary in characteristic yet different fashions.

In each of these three species, the maximal value of numerical density of the array of conventional synapses is limited. In chick and Xenopus, however, new conventional synapses are formed as the IPL grows. This is indicated by the increase in planimetric density. The rate of formation of these synapses is such that there is no increase in numerical density. These data suggest that the maximal numerical density of the array of conventional synapses is species specific and the number of synapses which can be present in a unit volume of neuropil is limited. If the neuropil grows in volume at a particular locus, new conventional synapses are formed. The planimetric density of the array is thus increased but without a concommitant increase in numerical density.

In mouse and in Xenopus, the planimetric densities of the arrays of ribbons are consistent with the hypothesis that the bipolar neurons form a certain number of ribbon synapses that does not change even as the volume of the neuropil grows. Since the arrays of the chick retina continue to grow after hatching (Morris, Wylie and Miles, 1977), the planimetric density of ribbons might follow the same pattern as in mouse and Xenopus.

Hypothesis for maintaining a constant numerical density:

Although no direct evidence can be offered in proof, a possible mechanism which would insure the constancy of numerical density even if the volume of neuropil changes is that each conventional synapse claims a territory of post-synaptic membrane upon which no other synapse can form. As growth of the array of conventional synapses proceeds, new synapses form on membrane as yet unclaimed. But, when the claimed territories begin to abut, no new synapses can be formed as there is no unclaimed membrane. The array thus exhibits a constant numerical density.

If the neuropil grows, more unclaimed membrane is added and thus new synapses can form, but while the formation of these new synapses increases the total number of synapses in the neuropil, the number per unit volume need not change. Indeed, as we have seen it does not change. The array therefore exhibits a constant numerical density with an increasing planimetric density. A competition for the available membrane would serve to limit the growth of the array in the sense of limiting the possibilities of synaptic interaction while preserving the ability of the neuropil to add synapses as its volume increases.

SUMMARY

1.   Intra- and inter- specific comparison of the growth curves for two synaptic sub-arrays determined by quantitative descriptors

has demonstrated that ribbon synapses and conventional synapses
are different in time of appearance and subsequent addition to the
neuropil.
2.   The presence of a primitive array formed by pioneering
conventional synapses before any possibility of a normal sensory
input suggests that the earliest synapses might be transmitting
information about the organization of the neuropil rather than
processing sensory data.
3.   Bipolar cells form connections in the inner and outer
plexiform layers simultaneously.
4.   There are species-specific values for numerical and
planimetric densities of arrays of synapses.

Descriptive statistics comparable to the array descriptors
presented here do not exist for any other portion of the
developing nervous system.   Other descriptors of the array can be
defined and specific sub-arrays can be investigated to further
refine our knowledge of the growth, development and functioning of
the retina.   The layered structure of the neuropils of other areas
of the central nervous system suggest that the concepts presented
here have a broad range of possible application.

REFERENCES

Blanks, J.C., A.M. Adinolfi, and R.N. Lolley, 1974.
    Synaptogenesis in the photoreceptor terminal of the mouse
    retina, J. Comp. Neurol., 156:81-94.
Boycott, B.B., J.E. Dowling, and S.K. Fisher, 1975.
    Interplexiform cells of the mammalian retina and their
    comparison with catecholamine-containing cells, Proc. R. Soc.
    Lond. (Biol. Sci.), 191:353.
Chen, F. and P. Witkovsky, 1978.   The formation of photoreceptor
    synapses in the retina of larval Xenopus, J. Neurocytol.,
    7:721-774.
Dowling, J.E. and B.B. Boycott, 1966.   Organization of the primate
    retina: electron microscopy, Proc. R. Soc. Lond. B.,
    166:80-111.
Dowling, J.E.   1968.   Synaptic organization of the frog retina: an
    electron microscopic analysis comparing the retinas of frogs
    and primates, Proc. R. Soc. Lond. B., 170:205-228.
Dubin, M.W. 1970.   The inner plexiform layer of the vertebrate
    retina: a quantitative and comparative electron microscopic
    analysis, J. Comp. Neurol., 140:479-506.
Fisher, L.J. 1976.   Synaptic arrays in the inner plexiform layer
    of the developing retina of Xenopus, Develop. Biol.,
    50:402-412.
Fisher, L.J. 1979a.   Development of synaptic arrays in the inner
    plexiform layer of neonatal mouse retina, J. Comp. Neurol.,
    187:359-372.
Fisher, L.J. 1979b.   Interplexiform cell of the mouse retina: a

Golgi demonstration, Invest. Ophthalmol., 18:521-523.

Fisher, L.J. and S.S. Easter, Jr. 1979. Retinal synaptic arrays: continuing development in the adult goldfish, J. Comp. Neurol., 185:373-380.

Fujita, S. and M. Horii, 1963. Analysis of cytogenesis in the chick retina by tritiated thymidine autoradiography,·Arch. Histol. Jap., 23:359-366.

Hamburger, V. and H. Hamilton, 1951. A series of normal stages in the development of the chick embryo, J. Morphol., 88:49-92.

Hollyfield, J.G. 1971. Differential growth of the neural retina in Xenopus laevis larvae, Develop. Biol. 24:264-286.

Kahn, A.J. 1974. An autoradiographic analysis of the time of appearance of neurons in the developing chick neural retina, Develop. Biol., 38:30-40.

McLaughlin, B.J. 1976. A fine structural and E-PTA study of photoreceptor synaptogenesis in the chick retina, J. Comp. Neurol., 170:347-364.

Meller, K. and W. Tetzlaff, 1976. Scanning electron microscopic studies on the development of the chick retina, Cell Tissue Res., 170:145-159.

Meller, K. and W. Tetzlaff, 1977. The development of membrane specializations in the receptor-bipolar-horizontal cell synapse of the chick embryo retina, Cell Tissue Res., 181:319-326.

Morris, V.B., C.C Wylie, and V.J. Miles, 1977. The growth of the chick retina after hatching, Anat. Rec., 184:111-114.

Nieuwkoop, P.D. and J. Faber, 1967. Normal Table of Xenopus laevis (Dandin), North Holland Publishing Co., Amsterdam.

Olney, J.W. 1968. An electron microscopic study of synapse formation, receptor outer segment development and other aspects of developing mouse retina, Invest. Ophthalmol., 7:250-268.

Olson, M.D. 1979. Scanning electron miscroscopy of developing photoreceptors in the chick retina, Anat. Rec., 193:423-438.

Raviola, G. and E. Raviola, 1967. Light and electron microscopic observations on the inner plexiform layer of the rabbit retina, Amer. J. Anat., 120:403-426.

Sidman, R.L. 1961. Histogenesis of the mouse retina studied with thymidine -^{3}H, in Smelser, G.K., ed., "The Structure of the Eye", Academic Press, 487-506.

Straznicky, K. and R.M. Gaze, 1971. The growth of the retina in Xenopus laevis: an autoradiographic study, J. Embryol. Exp. Morphol., 26:67-79.

Tucker, G.S. and J.G. Hollyfield, 1977. Modification by light of synaptic density in the inner plexiform layer of the toad, Xenopus laevis, Exp. Neurol., 55:133-151.

West, R.W. 1972. Thick sections of epon, Stain Technol., 47:201.

Physiological Development of Retinal Synapses

Ramon F. Dacheux

Studies which have examined the development of ganglion cell receptive fields emphasize the contrasting maturation sequence of center vs. surround organization. Initially, ganglion cell responses are dominated by the center, whereas the surround activity is either absent or very weak (Bowe-Anders et al., 1975; Rusoff and Dubin, 1977; Masland, 1977). Center-surround antagonism increases with age as the surround mechanism gradually matures and eventually achieves adult characteristics. In the rabbit this period of receptive field development occurs between 8 and 30 days of age: by 30 days of age most cells have adult receptive field properties.

In hopes of further increasing the understanding of the functional development of the retina and explaining the maturation of the surround response, an intracellular analysis of postreceptor neurons in the neonate rabbit retina was carried out. The research centers on: 1) the maturation rate of horizontal cell responses, since it is believed that horizontal cells mediate the surround activity of ganglion and bipolar cells; 2) the development of center-surround interactions in the bipolar cell responses to evaluate if they also display the same pattern of maturation as ganglion cells; 3) the development of amacrine cell responses which may also influence the shaping of ganglion cell receptive fields.

METHODS

An isolated retina eyecup preparation of the New Zealand White rabbit was used for all of the experiments. The in vitro retina eyecup circumvents most of the problems usually encountered when studying neonates, such as: poor optics of the eye and low tolerance to anesthesia. The rabbits were born in our animal facilities to observe the exact time of birth for age calculations. Retinas were examined between the ages of 8 days when ganglion cells first respond to light and 30 days when they display mature receptive field organizaton.

The eye was surgically enucleated under Nembutal anesthesia in dim red light. It was hemisected and the retina eyecup was everted and secured within a specially designed chamber, vitreous side up. A warm, bicarbonate buffered Ringer's solution that contained serum bathed the retina within three minutes of enucleation. The electroretinogram was constantly monitored to provide an evaluation of the retina's condition throughout the intracellular recording experiments. Microelectrodes, filled with 2 M potassium acetate, ranged between 100 to 1000 megohms in resistance. A hydraulic microdrive controlled the electrode as it penetrated the retina. The light stimuli (diffuse, small spot and annulus) were focused directly on the retinal surface and were presented as one second pulses. Intensity was varied by using neutral density filter wheels (Miller and Dacheux, 1973; Dacheux and Miller, 1981a).

Intracellular injections of Procion Yellow, used to locate and identify cellular responses, resulted in staining of only the cell body or a few isolated cell processes. Therefore, cell identification relied on the location of the stained cell body or processes within retinal layers and on properties such as: 1) relative retinal depth of the recording; 2) receptive field size; 3) presence of center-surround organization within the receptive field: and 4) presence or absence of spike activity. Intracellular responses recorded from rabbit retinal neurons did not differ from the responses reported in mudpuppy (Werblin and Dowling, 1969), fish (Kaneko, 1970) and cat (Nelson, 1977). In addition, recent intracellular recording and staining experiments (personal observation) have confirmed the identifications in the adult.

RESULTS AND DISCUSSION

An important and particularly fortunate finding is that immature neurons of the rabbit retina do not evolve from a "primitive" or "primordial" hyperpolarizing or depolarizing unit;

a few adult response characteristics are present when they first respond to light. Therefore, since it is possible to classify response types in the adult, it is possible to establish its neonate counterpart with similar features.

The results of this study suggest a sequence for functional maturation of the rabbit retina that is divided into three stages:

Stage I. a period of initial light sensitivity which starts at eight days of age and extends to ten days.

Stage II. a period of progressive acquisition of adult response features that occurs between days eleven and fifteen.

Stage III. a phase in which minor differences are apparent when compared to the adult which extends from sixteen days of age to thirty days.

Stage I (8-10 days of age)

During the first stage, in which postreceptor neurons acquire light responsiveness, horizontal cells respond to different intensities of stimulation with graded sustained hyperpolarizations (figure 1). Although the responses appear similar, certain properties separate them into two groups, originating from two different types of horizontal cells: axonless and axon-bearing cells. The first difference is seen in the amplitude-intensity function; the responses from the soma of the axon-bearing cell needs one log unit less light intensity than the axonless cell to obtain half of the maximum response amplitude. Secondly, a secondary hyperpolarization is consistently observed at the termination of intense levels of illumination. Chromatic adaptation equated for equal rod function demonstrates that this after potential seen in rabbit horizontal cell responses (De Monesterio, 1978) is similar to the rod after-effect first described in the cat by Steinberg (1969) and more recently by Nelson (1977). The classification of rabbit horizontal cell responses relies on the evaluation of the rod-cone contribution based on the relative amplitude of fast and slow response components observed at the termination of illumination (Dacheux and Miller, 1981a). The waveforms are similar but the responses from the soma of the axon-bearing cell have a larger rod after-effect, especially at the higher intensities. Area of summation is the third property to separate the horizontal cell responses into two groups. The axonless cells sum over an area 500 μm in diameter, whereas the summation area of the responses from the soma of the axon-bearing cells is less.

Responses from depolarizing and hyperpolarizing bipolar cells are characterized by sustained monophasic potentials (figure 1) which return immediately to the base line at the termination of low to moderate levels of diffuse light stimulation. Stimulation of depolarizing bipolar cells with moderate to high intensity light pulses results in a small, slow, secondary depolarizing after-potential; by analogy with the horizontal cell after-effect this potential may also reflect rod activity. However, it is impossible to verify this assumption by using chromatic adaptation because of the difficulty in maintaining the intercellular recordings for long periods of time. No after-effect is observed for hyperpolarizing bipolars. A 700 µm diameter spot evokes a maximum response amplitude in depolarizing bipolar cells: attenuation of the responses occurs with smaller diameter stimuli. Hyperpolarizing bipolar cell recordings are difficult to maintain, so accurate receptive field measurements are not available. These monophasic responses from depolarizing bipolars, therefore, sum over large areas and have no antagonistic surround at this stage of their development.

Figure 1. A schematic illustration of a simplified retinal network during "Stage I" (8-10 days of age) of development, to demonstrate: 1) the basic structural characteristics associated with synaptic development; and 2) the typical characterisics of the intracellular responses obtained from the immature neurons. The diagram illustrates primary contact, in the OPL, of horizontal (H) and bipolar (B) cell processes with a non-invaginated photoreceptor (R) terminal containing a sparse population of vesicles. Processes in the IPL are also initiating contact as amacrine (A) and ganglion (G) cell dendrites form synapses with bipolar cell terminals which also contain a limited number of vesicles. Seen in the top right are two different horizontal cell recordings; one from an axonless cell and the other from the soma of the axon-bearing cell. The main difference in these responses is the varying amount of rod after-effect (arrows). The middle responses are from two different bipolar cells; a depolarizing and hyperpolarizing cell. These recordings demonstrate the monophasic nature of the responses. The recording from an immature "on-off" amacrine cell is seen at the bottom right illustrating the presence of only an "on" depolarization and the absence of an "off" response. All potentials are responses to diffuse illumination; irradiance is 1.0×10^{-5} W/cm^2. The dark bars under the responses indicate the duration of the stimulus. Calibrations for each response are marked.

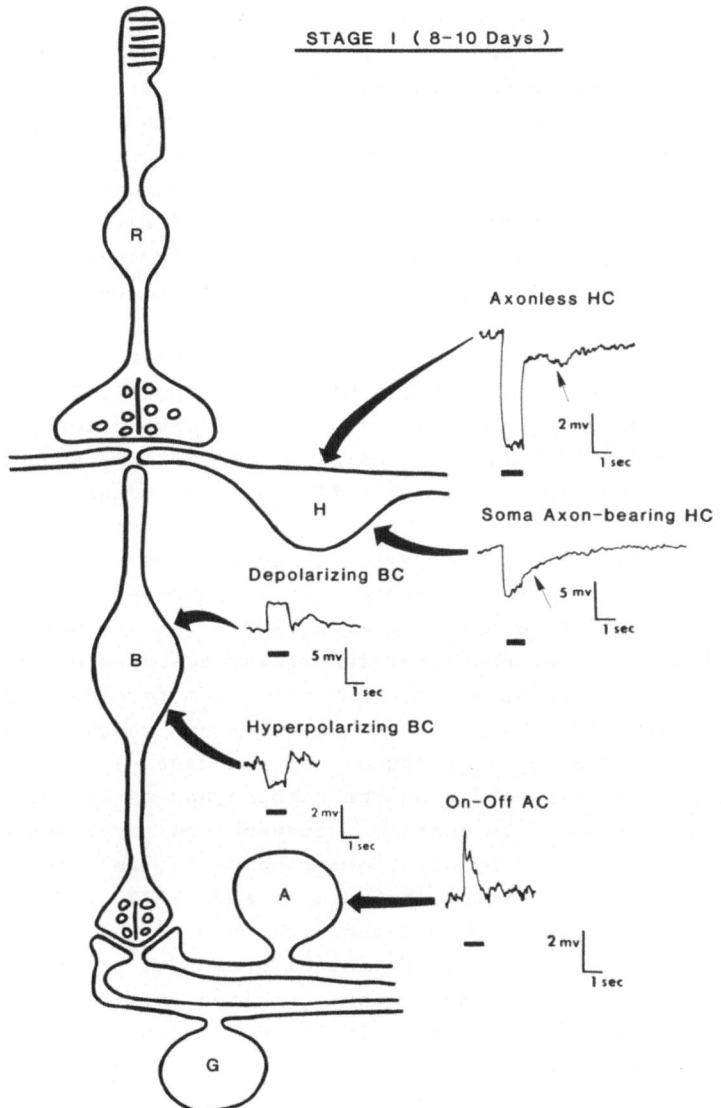

STAGE I (8-10 Days)

Axonless HC

Soma Axon-bearing HC

Depolarizing BC

Hyperpolarizing BC

On-Off AC

2 mv
1 sec

5 mv
1 sec

5 mv
1 sec

2 mv
1 sec

2 mv
1 sec

R

H

B

A

G

Immature "on-off" amacrine cell responses usually appear as a transient depolarization at the onset of stimulus illumination with either no "off" depolarization or a poorly developed one (figure 1). It is impossible to elicit an "off" response in some of these cells although various patterns of stimuli and different levels of light adaptation are employed. Immature amacrine cells depolarize to a peak with a fairly fast rise time and almost immediately repolarize to the baseline before the termination of a one second stimulus pulse. Therefore, the depolarization of "on-off" amacrine cells is different than the monophasic sustained depolarization associated with "on" bipolar cells (Dacheux and Miller, 1981b). Towards the end of this developmental stage most "on-off" amacrine cells display weak "off" responses which are somewhat labile.

At eight days of age, the outer segments of rabbit photoreceptors are 15% of their adult length (Noell, 1958). Their synaptic terminals are small and contain a sparse population of synaptic vesicles, usually in a halo surrounding the synaptic ribbon. The immature synapse is a flat contact occurring at the base of the terminal, adjacent to a synaptic ribbon; it consists of a dendrite from a bipolar cell flanked on each side by a horizontal cell process. No invaginating synapses are observed (Blanks et al., 1974; McArdle et al., 1977). At the same time, synaptogenesis in the inner plexiform layer has been progressing for only two or three days. Bipolar cell axon terminals have just started to form dyad synapses with amacrine and ganglion cells at sites adjacent to synaptic ribbons; vesicle density is sparse and restricted to an area bordering the ribbons and presynaptic membrane thickenings. In contrast, conventional synapses similar to serial synapses made by adult amacrine cells are observed much earlier than ribbon synapses (McArdle et al., 1977). The completion of this synaptogenic sequence in both the outer and inner plexiform layers marks the establishment of primary function in horizontal, bipolar, amacrine and ganglion cells. Therefore, with the morphology and function in place, elementary information processing appears to be initiated within the major radially conducting neuronal network of the retina, and is expressed at its output by simple ganglion cell responses.

Stage II (11-15 days of age)

Between eleven and fifteen days of age the retinal neurons

enter a transitional stage of synaptic function; dramatic changes occur in the waveform of the light responses as they begin to achieve adult appearances. Coincident with the initiation of this developmental stage is the time of first eye opening.

At this intermediate maturational stage, the axonless horizontal cells start to display a small transient component in response to the onset of high level pulses of illumination (figure 2). With age this response component gradually increases in size to become a prominent feature of the mature response; transient potentials are never observed before eleven days of age. A dynamic increase in response amplitude is also observed during this period as it increases to more than half of the mature amplitude. This increase in response amplitude makes the rod after-effect appear reduced in size although it remains relatively constant throughout maturation. The responses recorded from the soma of the axon-bearing horizontal cell differ from those of the axonless cell by displaying an increase in the amplitude and duration of the rod after-effect (figure 2). Therefore, the rod after-effect becomes a prominent feature of these responses and establishes a demarcation between them and the responses from axonless horizontal cells; a small transient component is frequently observed but it never develops into a major response characteristic like the axonless cell recordings. Similar to the axonless horizontal cell responses, however, these reponses also increase substantially in their amplitude to light. Both types of responses have areas of summation which are 50% smaller than observed in the fully mature state: the soma of the axon-bearing cells sum over an area which never exceeds 500 μm in diameter; axonless cells sum over an area between 700 and 900 μm in diameter.

At this stage, horizontal cell responses from the axon terminal of the axon-bearing horizontal cell are first observed and are characterized by a small, relatively sustained hyperpolarization during illumination and a large dominant rod after-effect at the termination of high intensity stimulation; no evidence of a transient component is ever observed (figure 2). The amplitude-intensity function is broader than in the other horizontal cells with a threshold intensity half a log unit lower. These cells sum over a small area which at this age never exceeds a diameter of 400 μm.

During the early part of this stage, bipolar cell responses

start to show a suggestion of transient and sustained light evoked
activity (figure 2). In depolarizing bipolar cells the
after-potential is evident at intensities two log units above
threshold. Hyperpolarizing bipolar cells do not have a secondary
hyperpolarization or after-effect; the presence or absence of this
response component is one way of differentiating horizontal from
hyperpolarizing biploar cell responses. With the first appearance
of the transient response component, a test for antagonistic
center-surround organization does not reveal surround activity.
Later in this period when prominent transient and sustained
waveforms appear, additional organizational characteristics of the
receptive field are apparent: large diameter stimuli reduce the
response amplitude and enhance the phasic component; a focal
stimulus of 100 μm in diameter produces the maximum monophasic
response; presentation of an annulus with 100 μm inside and 240 μm
outside diameters elicits a response similar to focal stimulation
and not the expected antagonistic surround response of an opposite
polarity. By superimposing flashes of annular stimulation on a
continuous 100 μm diameter spot of illumination positioned in the

Figure 2. A schematic illustration of a simplified retinal
network during "Stage II" (11-15 days of age) of development which
demonstrates: 1) structural characteristics associated with
synaptic development; 2) response characteristics from maturing
neurons. The diagram illustrates an intermediate stage of
synaptic formation in the OPL as horizontal (H) and bipolar (B)
cell processes partially move into a slightly invaginated
photoreceptor (R) terminal containing a moderate population of
vesicles. In the IPL, the amacrine (A) and ganglion (G) cell
dendrites have established synapses with bipolar cell axon
terminals containing less than the normal number of vesicles.
Seen in the top right are three horizontal cell recordings from
three horizontal cell structures: the axonless cell, the soma of
the axon-bearing cell and the axon terminal of the axon-bearing
cell (see text for explanation). The main difference in these
responses is the amount of rod after-effect each response displays
(arrows). The middle left responses are from two different
bipolar cells: the depolarizing and hyperpolarizing cells. The
recordings on the left have a small transient component associated
with a sustained potential; furthermore, the recordings on the
right establish the presence of antagonistic center-surround
receptive field organization for the first time during this stage
of maturation. A recording from an "on-off" amacrine cell is seen
at the bottom and illustrates the slow transient depolarizations
in response to the initiation and termination of the stimulus.
All potentials are responses to diffuse illumination; irradiance
is 1.0×10^{-5} W/cm^2. The dark bars under the response indicate
the presentation of the stimulus; ANN indicates annulus.
Calibrations for each response are marked.

STAGE II (11–15 Days)

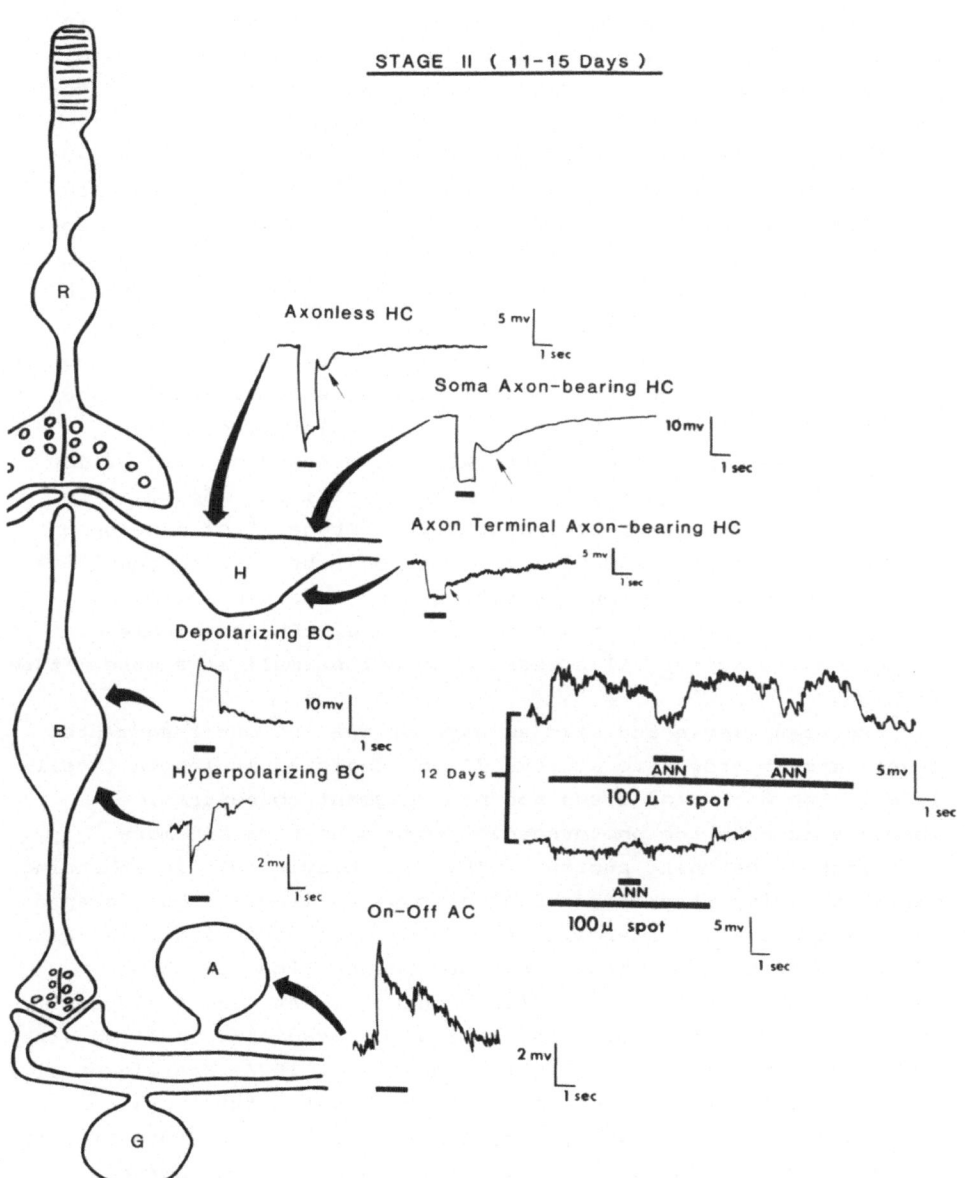

Axonless HC

5 mv
1 sec

Soma Axon-bearing HC

10 mv
1 sec

Axon Terminal Axon-bearing HC

5 mv
1 sec

Depolarizing BC

10 mv
1 sec

Hyperpolarizing BC

12 Days

100 μ spot

ANN   ANN

5 mv
1 sec

2 mv
1 sec

ANN

100 μ spot

5 mv
1 sec

On-Off AC

2 mv
1 sec

R

H

B

A

G

center, the antagonistic surround response consisting of a
monophasic potential opposite in polarity to the center response
is elicited (figure 2). This demonstrates that center-surround
organization of the bipolar cells has indeed reached a certain
state of maturity; furthermore, very few sustained monophasic
bipolar cell responses to diffuse illumination are observed during
the latter part of this developmental stage.

Amacrine cell responses during the first part of this period
have relatively large "on" responses and labile "off" responses.
The "off" responses are usually more prominent when large diameter
or long duration stimuli are used (figure 2). Distinct transient
depolarizations are observed in responses to the initiation and
termination of light stimuli with diameter 700 μm or larger. If
the intensity and duration of the stimuli remain constant, and its
diameter is decreased, the "off" response is considerably
diminished until it is obscured in the repolarizing phase of the
"on" response. Responses evoked with stimulus diameters of 200 μm
or less appear as transient "on" depolarizations, identical to
amacrine cell responses recorded during "Stage I" of development
(Dacheux and Miller, 1981b). In addition, the use of repetitive
stimuli of short duration flashes results in a diminution and
periodic elimination of the "off"-response. By the middle of this
period most amacrine cells respond to all stimuli in a more mature
fashion.

Between eleven and sixteen days of age the outer segments
accelerate in growth to achieve 70% of their adult length (Noell,
1958). The number of discs and photopigment concentration
increase to make the photoreceptor more sensitive and more
efficient in catching photons. This may play a role in enhancing
the sensitivity of photoreceptor neurons to light. Photoreceptor
terminals are also undergoing secondary changes during this stage
of development. The initial synaptic connections at these
terminals are established as flat contacts. At ten or eleven
days, the presynaptic membrane starts to invaginate, pulling with
it the postsynaptic elements (Blanks et al., 1974; McArdle et al.,
1977). Little is known of the minute ultrastructural changes
associated with this event; they probably result in a modification
of synaptic function by: (i) providing a site of synaptic
interaction between photoreceptors and second order neurons,
and/or between two different second order neurons; (ii) increasing
the efficiency of transmitter action to enhance postreceptor cell

sensitivity. This time period of secondary synaptogenic
modification is coincident with the first appearance and
maturation of surround responses in bipolar cells and the enhanced
sensitivity of horizontal cell responses. Although modifications
in the morphology of photoreceptor terminals can be correlated
with functional maturation of bipolar cell surrounds and
horizontal cell development, no horizontal cell feedback onto the
photoreceptor terminal to mediate the biplolar cell surround
response has ever been demonstrated in mammals; a correlation is
far from conclusive evidence. Maturation of synapses in the inner
plexiform layer involves a progressive increase in the density of
synaptic vesicles, particularly near the presynaptic membrane
thickening and synaptic ribbon (McArdle et al., 1977). Therefore,
as the number of vesicles per unit area of synaptic contact
increases, so does the functional efficiency of the synapse in
transmitting an electrical signal. Perhaps the synaptic fatigue
observed in the "off" response of immature amacrine cells is the
result of a low density of synaptic vesicles per synapse which in
turn leads to an inability to maintain vesicles near the membrane
release site during small spot diameter or repetitive stimulation.
In addition, since "on-off" amacrine cells receive excitatory
input from depolarizing bipolar cells at initiation of light and
excitatory input from hyperpolarizing bipolars at termination of
light (Miller and Dacheux, 1976a; Dacheux et al., 1979), the
failure to detect an "off" response or a weak "off" response in
amacrine cells suggests that the axon terminals of the two
different bipolar cells are not maturing at the same rate.

Stage III (16-30 days)

It is apparent that most intracellular responses recorded at
the start of this age period are very similar to their counterpart
in the adult and obtain maturity by the end of it. Horizontal
cell responses are of three distinct types (figure 3). Axonless
horizontal cell responses increase in amplitude and finally result
in large potentials that sometimes reach 50 mv; a predominant
transient component and small rod after-effect characterize
responses to diffuse illumination of high intensity. The response
of the soma of the axon-bearing horizontal cells, however, does
not increase very much in amplitude during this stage and
responses to high intensity diffuse light stimulation generally
never exceed 30 mv. The waveform is relatively sustained with a

hint of a transient component and a prominent rod after-effect
that usually reaches an amplitude of 50% of the primary
hyperpolarization. On the other hand, the axon terminal of the
axon-bearing horizontal cell responds to high levels of
illumination with a sustained hyperpolarization of 20 mv or less;
the rod after-effect can equal the initial light response and take
several seconds to repolarize back to baseline. When the area of
summation of these three responses is compared: the axonless
response sums over an area 1.5 to 2.0 mm in diameter, the soma of
the axon-bearing cell responds over an area that does not exceed 1
mm in diameter; and the axon terminal of the axon bearing cell has
a summation area of 500 μm in diameter. Therefore, the area of
summation for the soma of the axon-bearing cells and the axonless
cells has increased substantially from the preceding stage of
development; the axon terminal of the axon-bearing cell has not
changed much in its summation properties.

Depolarizing and hyperpolarizing bipolar cell responses to
diffuse illumination during this period consist of biphasic
potentials in which an initial transient declines to a sustained
plateau phase and repolarizes back to baseline at the stimulus
termination (figure 3). Focal stimulation of the receptive field
center elicits a monophasic potential of the same polarity as
diffuse illumination. It is not possible to evoke a pure surround

Figure 3. A schematic illustration of a simplified retinal
network during "Stage III" (16-30 days of age) of development
which demonstrates synaptic structural characteristics and
response characteristics of young mature neurons. The diagram
illustrates a mature synapse in the OPL as horizontal (H) and
bipolar (B) cells occupy the recess of a fully invaginated
photoreceptor (R) terminal filled with vesicles. In the IPL,
amacrine (A) and ganglion (G) cell dendrites synapse with a
bipolar cell axon terminal also filled with vesicles. Seen at the
top right are three different mature horizontal cell responses
recorded from the axonless cell, the soma of the axon bearing cell
and the axon terminal of the axon bearing cell (see text). The
difference in these responses is seen in the size and duration of
the rod after-effect (arrows). The middle recordings are from two
different cells: the depolarizing bipolar cell and the
hyperpolarizing bipolar cell. These recordings demontrate the
transient-sustained nature of the mature responses. The mature
"on-off" amacrine cell response is seen at the lower right and
consists of transient depolarizations at the onset and offset of
stimulus illumination. All potentials are responses to diffuse
light stimulation; irradiance is 1.0×10^{-5} W/cm^2. The dark bars
under the responses represent the presentation of the stimulus.
Calibrations for each response are marked.

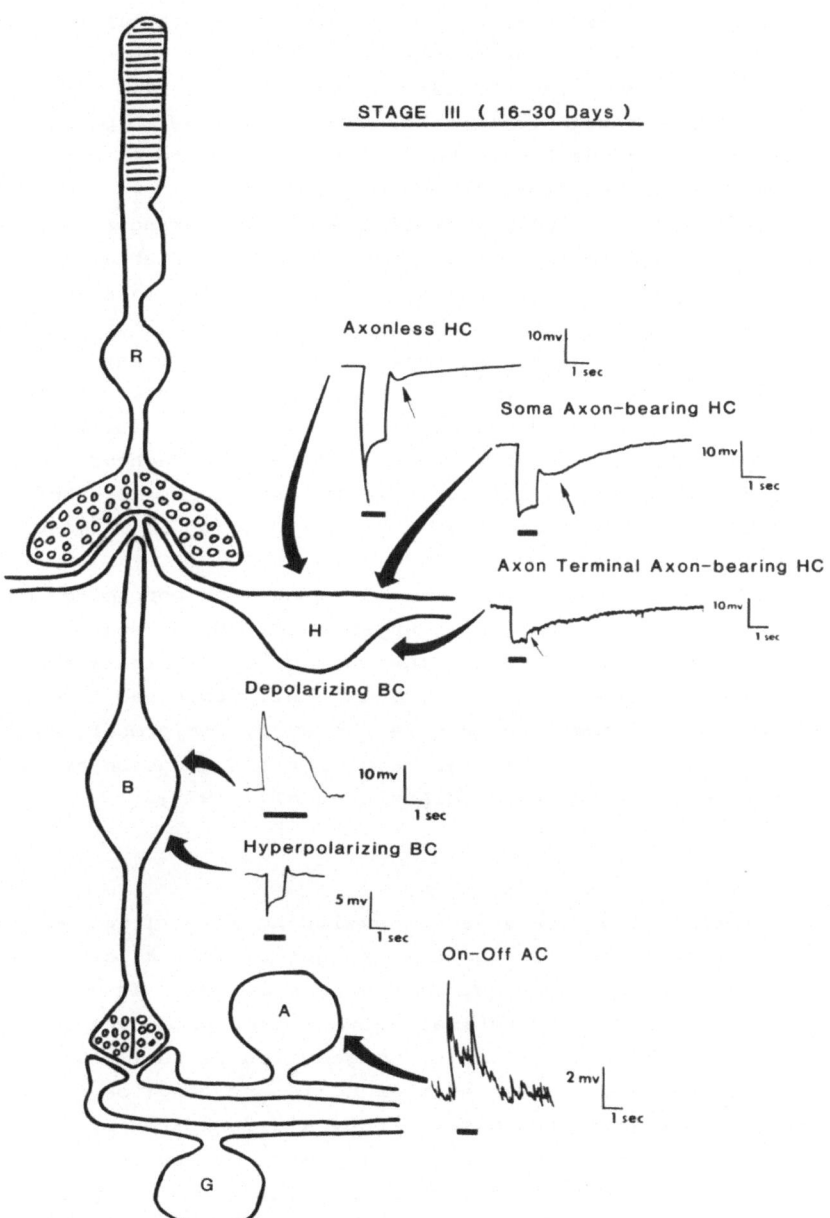

STAGE III (16-30 Days)

Axonless HC

10 mv

1 sec

Soma Axon-bearing HC

10 mv

1 sec

Axon Terminal Axon-bearing HC

10 mv

1 sec

Depolarizing BC

10 mv

1 sec

Hyperpolarizing BC

5 mv

1 sec

On-Off AC

2 mv

1 sec

response by stimulation with only an annulus; the annulus must be superimposed on continuous adapting focal illumination to suppress the dominating influence from the center. Thus, rabbit bipolar cells have receptive fields organized into a center with an antagonistic surround but similar to mudpuppy (Werblin and Dowling, 1969) and tiger salamander (Werblin, 1978) bipolar cells, it is necessary to adapt out the center response to elicit the antagonistic response from the surround.

"On-off" amacrine cells show little change during this period and appear similar to those described in lower vertebrates (Werblin and Dowling, 1969). They respond with a transient depolarization at the onset and termination of the light stimulus (figure 3). Toward the end of this stage two different size spikes, large somatic and smaller dendritic ones, are observed riding on the transient depolarizations, a property initially described in mudpuppy amacrine cells (Miller and Dacheux, 1976b). In addition, some cells display a slight enhancement of their response amplitude when stimulated with 100 to 50 μm diameter spots of light.

By the end of this stage of development, the morphology of the retina is unmistakably mature in appearance. Outer segments of the photoreceptors are at adult length and the terminals are fully invaginated with bipolar and horizontal cell processes forming a typical triad. In the inner retina the dyad, reciprocal, and conventional synapses of bipolar, amacrine and ganglion cells appear in their mature state (McArdle et al., 1977).

SUMMARY

An intracellular analysis of developing retinal neurons of the rabbit reveals three sequential stages of maturation. The first stage extends from eight to ten days of age. In essence the major radially conducting retinal network from photoreceptor terminals, through bipolar cells, to ganglion cells is functional on a simplified level. When bipolar cells and ganglion cells initially become light responsive, they are dominated by their center responses and do not exhibit any surround activity: lateral interactions through horizontal and amacrine cells are probably non-existent as a result of the immaturity of their responses. If lateral interactions mediate surround influences as others suggest (Werblin and Dowling, 1969; Kaneko, 1970; Richter and Simon, 1975;

Naka, 1976; Miller and Dacheux, 1976a; Masland, 1977), then the
morphological site is either not formed or non-functional. The
second and most critical stage, which covers eleven to fifteen
days of age, is signaled by the initiation of eye opening and
dramatic modifications in all of the light responses of retinal
neurons. The antagonistic center-surround characteristics appear
in the receptive field organization of ganglion cells (Bowe-Anders
et al., 1975; Rusoff and Dubin, 1977; Masland, 1977) and bipolar
cells (Dacheux and Miller, 1981a). These lateral interactions
coincide with the gradual transition of horizontal and amacrine
cells to maturity, as well as critical morphological alterations.
By the third stage of development, sixteen to thirty days of age,
most responses from retinal neurons have achieved adult waveforms
and properties. One important difference is seen in the receptive
field organization of bipolar cells and ganglion cells. Focal
adaptation of the center is necessary to elicit an isolated
surround response from juvenile rabbit bipolar cells. Conversely,
at the same age, ganglion cell surrounds are evoked without any
center adaptation and annular stimulation alone. Therefore,
horizontal cells probably contribute to the antagonistic surround
mechanism of bipolar cells through lateral interaction in the
outer plexiform layer. On the other hand, amacrine cells may be
responsible for modifying the antagonistic surround activity of
ganglion cells through lateral interactions in the inner plexiform
layer.

ACKNOWLEDGMENTS

The expert photographic assistance of Mr. Peter Ley and Ms.
Suzanne Kuffler is gratefully acknowledged. This research was
supported by USPH grant EY-0084.

REFERENCES

Blanks, J.C., A. Adinolfi and R. Lolley. 1974. Synaptogenesis in
    the photoreceptor terminal of the mouse retina. J. Comp.
    Neurol.156: 81-93.
Bowe-Anders, C., R.F. Miller and R.F. Dacheux. 1975. Developmental
    characteristics of receptive organization in the isolated
    retina eye-cup of the rabbit. Brain Res. 87: 61-65.
Dacheux, R.F., T.E. Frumkes and R.F. Miller. 1979. Pathways and
    polarities of synaptic interactions in the inner retina of the
    mudpuppy: Synaptic blocking studies. Brain Res. 161: 1-12.
Dacheux, R.F. and R.F. Miller. 1981a. An intracellular
    electrophysiological study of the ontogeny of functional
    synapses in the rabbit retina. I. Receptors, horizontal and

46
bipolar cells. J. Comp. Neurol. 198: 301-326.

Dacheux, R.F. and R.F. Miller. 1981b. An intracellular electrophysiological study of the ontogeny of functional synapses in the rabbit retina. II. Receptors, horizontal and bipolar cells. J. Comp. Neurol. 198: 327-334.

DeMonasterio, F.M. 1978. Spectral interactions in horizontal and ganglion cells of the isolated and arterially perfused rabbit retina. Brain Res. 150: 239-258.

Kaneko, A. 1970. Physiological and morphological identification of horizontal, bipolar and amacrine cells in goldfish retina. J. Physiol. (Lond.) 207: 623-633.

Masland, R.H. 1977. Maturation of function in developing rabbit retina. J. Comp. Neurol. 175: 275-286.

McArdle, C.B., J.E. Dowling and R.H. Masland. 1977. Development of outer segments and synapses in the rabbit retina. J. Comp. Neurol. 175: 253-274.

Miller, R.F. and R.F. Dacheux. 1973. Information processing in the retina: Importance of chloride ions. Science 181: 266-268.

Miller, R.F. and R.F. Dacheux. 1976a. Synaptic organization and ionic basis of "on" and "off" channels in the mudpuppy retina. III. A model of ganglion cell receptive field organization based on chloride free experiments. J. Gen. Physiol. 67: 679-690.

Miller, R.F. and R.F. Dacheux. 1976b. Dendritic and somatic spikes in mudpuppy amacrine cells: Identification and TTX sensitivity. Brain Res. 104: 157-162.

Naka, K.I. 1976. Neuronal circuitry in the catfish retina. Invest. Ophthalmol. 15: 926-935.

Nelson, R. 1977. Cat cones have rod input: A comparison of the response properties of cones and horizontal cell bodies in the retina of the cat. J. Comp. Neurol. 172: 109-136.

Noell, W.K. 1958. Differentiation, metabolic organization and viability of the visual cell. Arch. Ophthalmol. 60: 702-733.

Richter, A. and E.J. Simon. 1975. Properties of center hyperpolarizing, red-sensitive biopolar cells in the turtle retina. J. Physiol. (Lond.) 248: 317-334.

Rusoff, A. and M. Dubin. 1977. Development of receptive field properties of retinal ganglion cells in kitten. J. Neurophysiol. 40: 1188-1198.

Steinberg, R.H. 1969. The rod after-effect in S-potentials from the cat retina. Vision Res. 9: 1345-1355.

Werblin, F.S. 1977. Synaptic interactions mediating bipolar responses in the retina of the tiger salamander. In "Vertebrate Photoreception" (Barlow and Fatt, eds.) London: Academic Press, pp. 205-230.

Werblin, F.S. and J.E. Dowling. 1969. Organization of the retina of the mudpuppy Necturus maculosus: II. Intracellular recording. J. Neurophysiol. 32: 339-355.

Development of Cholinergic and Amino Acid Neurotransmitter Systems in the Chick Retina

Robert Baughman

The presence of a cholinergic system in many vertebrate retinas is well established from biochemical and physiological experiments (Lindeman, 1947; Hebb, 1957; Graham, 1974; Ross and McDougal, 1976; Masland and Ames, 1976; Baughman and Bader, 1977). The first cellular localization of this system was achieved by the use of an autoradiographic technique with chick retina (Baughman and Bader, 1977). A similar technique was applied with similar results in the rabbit retina (Masland and Mills, 1979).

We were interested in the question of how the cholinergic system develops during embryogenesis. Lindeman (1947) observed a sharp increase in the endogenous content of acetylcholine (ACh) in the chick retina at about the time of hatching. We were surprised to find that the activity of the synthetic enzyme for ACh, choline acetyltransferase, changed only slightly during this period (Figure 1; Bader, et al., 1978). Synthesis and storage of [^{3}H]ACh from [^{3}H] choline, however did exhibit a sharp increase just before hatching (Figure 3; Bader, et al., 1978). The explanation for these results appears to be that there is an increase just before hatching in the ability of the tissue to accumulate [^{3}H]choline from the medium via a high-affinity transport mechanism (Table 3; Bader, et al., 1978). Thus in the chick retina the cholinergic system appears to develop in two stages. A large increase in choline acetyltransferase occurs by day 11, soon after cell division ends and well before synaptogenesis; and an increase in endogenous ACh content, high-affinity choline uptake and synthesis and storage of ACh occurs just before hatching, when synaptogenesis is most active. An important conclusion that can be drawn from these results is that the high-affinity uptake system for choline is not inevitably linked to acetylation by choline acetyltransferase.

Given the apparent two-stage maturation of the cholinergic
system in chick retina, we were interested to look at the time
course of appearance of putative amino acid neurotransmitters. We
also wished to survey as broad a series of amino acids as possible
in an effort to detect any changes in endogenous content that
might point out heretofore unsuspected neurotransmitters.
Developmental changes in amino acid content of chick retina have
been studied before (Pasantes-Morales, et al., 1973), but this
work did not include the period before day 10, where we know from
our cholinergic study that important changes can occur. We have
also monitored some additional compounds. A high pressure liquid
chromatographic separation, employing a cation exchange column and
post-column derivatization with o-phalaldehyde (Baughman and
Gilbert, 1981) was used to monitor the retinal content of 19
primary-amino-containing substances (the functional group detected
by the o-phalaldehyde reagent) from embryonic day 7 through
hatching at day 21. A histogram of the substances measured is
shown in Figure 1, with the levels present at day 21. From the
differences in the content observed at day 7 (Figure 2), it is
clear that several compounds undergo substantial increases in
endogenous content during retinogenesis. This is indicated more
clearly in Figure 3 where the ratio of the content at day 21 to
that of day 7 is shown. Glycine, β-alanine, γ-aminobutyric acid
(Gaba) and an unknown substance were particularly notable. The
details of the time course for these compounds, measured at
two-day intervals and plotted relative to the concentration at day
7, are shown in figures 4 and 5. Although these compounds begin
to increase by day 11, in each case most of the increase occurs
later. The values for 14 other compounds are presented in Figure
6.

The identification of the unknown compound plotted as peak
17, which underwent more than a one hundred-fold increase in
concentration, was clearly of importance. Hydrolysis of the
retinal extract with 6N hydrochloric acid was used to determine
whether the unknown was a peptide or an amino acid. Following
hydrolysis the unknown substance disappeared, β-alanine increased
greatly and a previously undetected substance comigrating with
1-methylhistidine appeared. This strongly suggests that the
unknown compound may be a peptide containing β-alanine and
1-methylhistidine. A likely candidate is anserine (β-alanayl-1

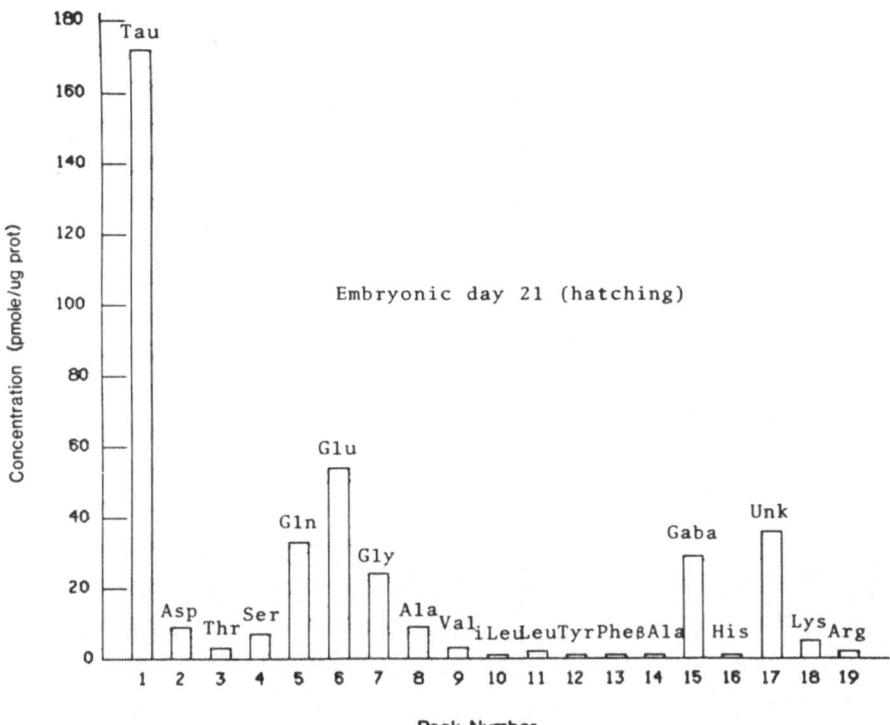

Figure 1. Endogenous primary-amine-containing compounds present in chick retina on embryonic day 21 (hatching). Analyses were performed on 1 N perchloric acid extracts (10:1 volume/wet weight). This and all following determinations were made in quadruplicate. The standard deviations ranged from 10-30%. Tau: taurine, Asp: aspartate, Thr: threonine, Ser: serine, Gln: glutamine, Glu: glutamate, Gly: glycine, Ala: alanine, Val: valine, iLeu: iso-leucine, Leu: leucine, Tyr: tyrosine, Phe: phenylalanine, βAla: β-alanine, Gaba: γ-aminobutyric acid, His: histidine, Unk: unknown, tentatively identified as anserine, Lys: lysine, Arg: arginine.

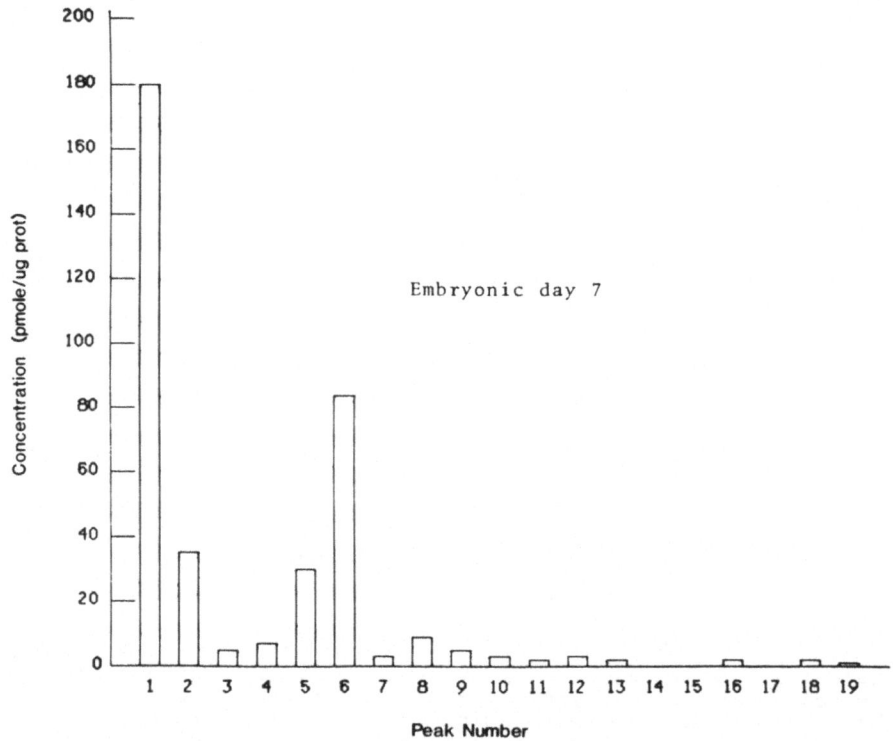

Figure 2. Endogenous primary-amino-containing compounds present
in chick retina at embryonic day 7. Conditions and peak number
identifications as in figure 1.

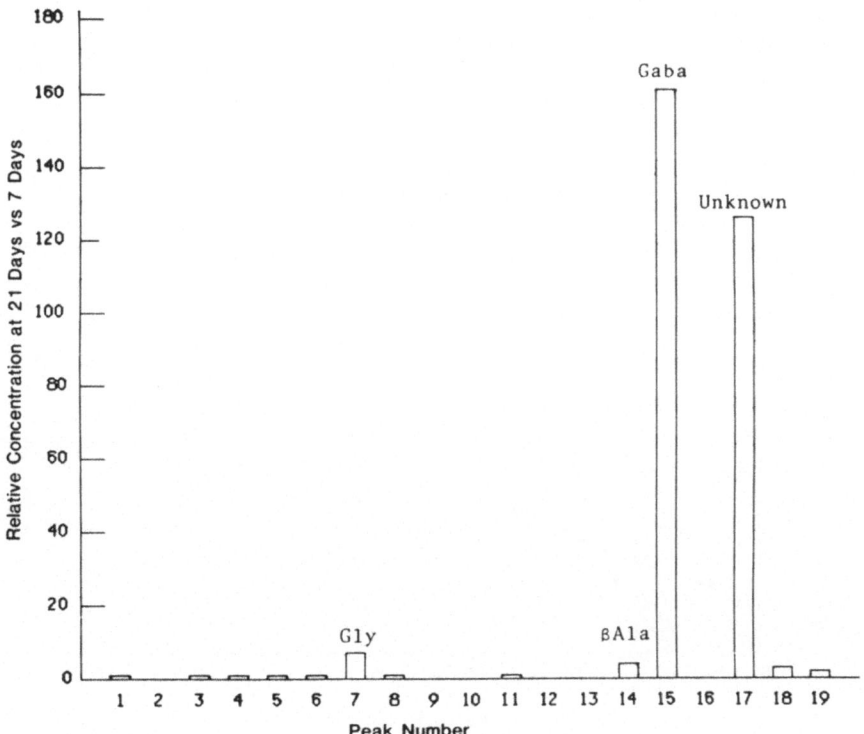

Figure 3. Relative concentration of endogenous compounds at embryonic day 21 versus embryonic day 7 (ratio of the values shown in figure 1 and 2).

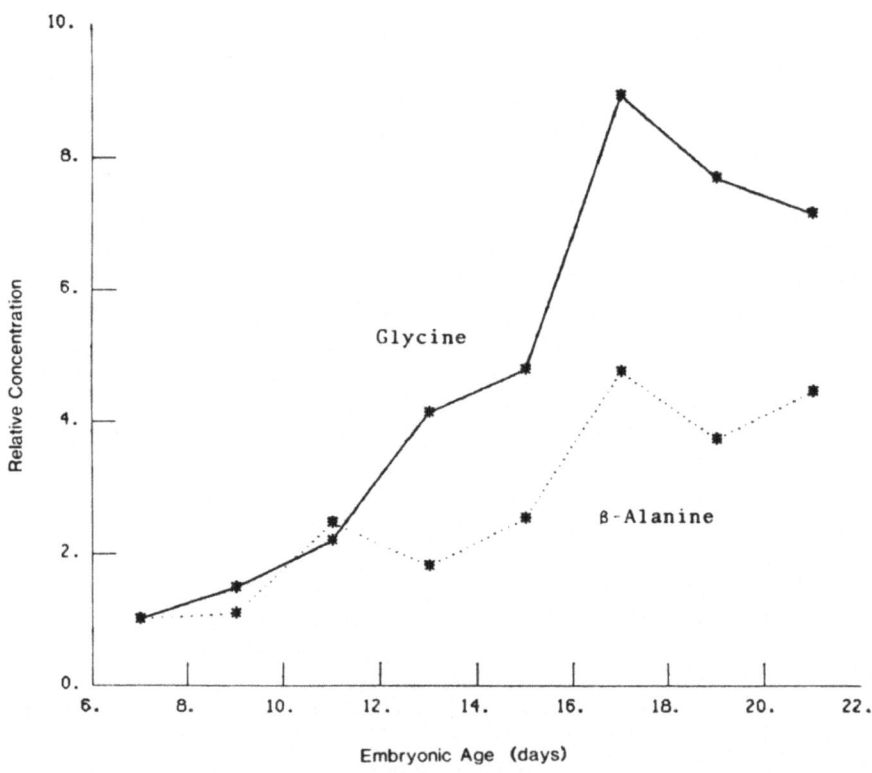

Figure 4. Concentration of glycine and β-alanine in chick retina from embryonic day 7 to embryonic day 21 (hatching) expressed relative to their concentration at day 7. Conditions as in figure 1.

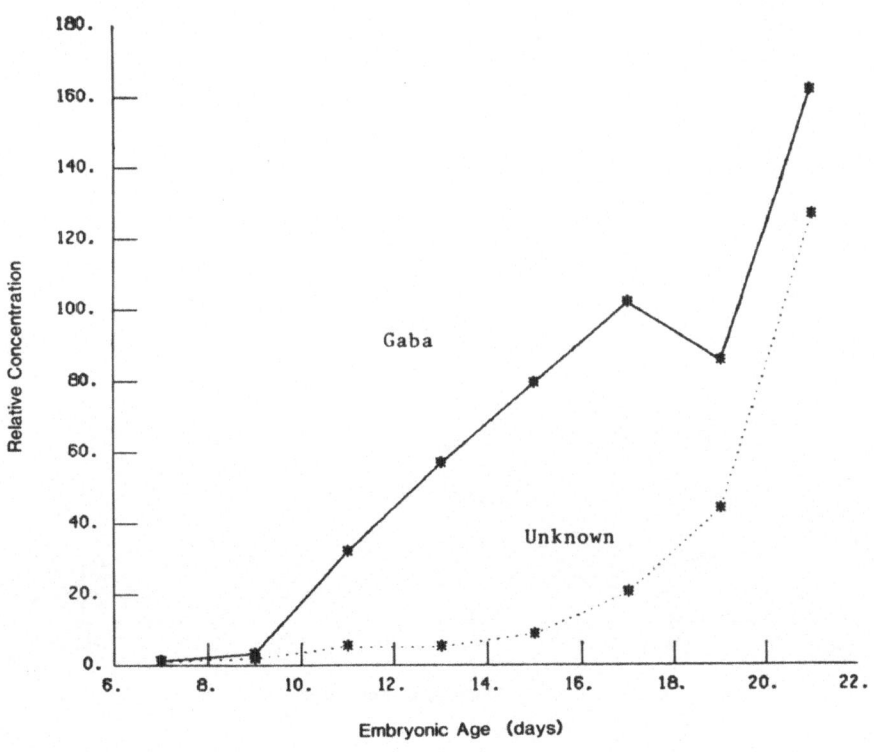

Figure 5. Concentrations of Gaba and an unknown compound (tentatively identified as anserine) in chick retina from embryonic day 7 to embryonic day 21 (hatching) expressed relative to their concentration at day 7. Conditions as in figure 1.

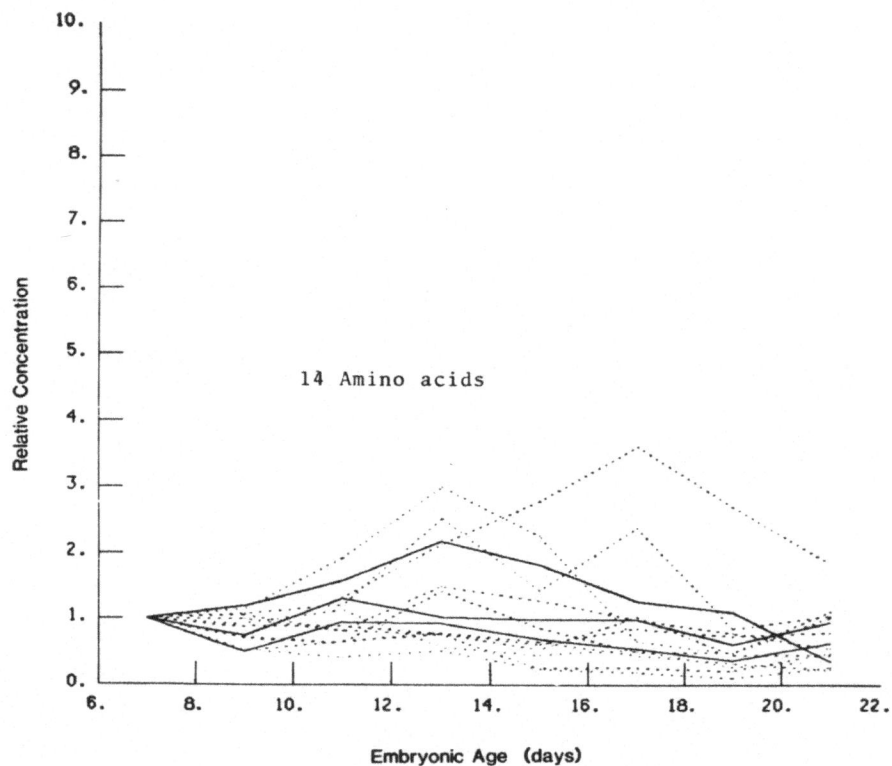

Figure 6. Concentration of amino acids other than those shown in figures 4 and 5 from embryonic day 7 to embryonic day 21 (hatching) expressed relative to their concentration at day 7. Conditions as in figure 1.

methylhistidine), a compound previously reported to be present in chick brain (Fisher et al., 1977). In fact, an authentic sample of this substance comigrated with the unknown peak. Extracts were prepared from retinas of other species to determine how general the presence of anserine was. A high concentration was present in turtle(Pseudymys elegans), but in goldfish and rat retinas almost none was observed.

In summary, the cholinergic system in the chick retina develops in two stages; a more than 100 fold increase in choline acetyltransferase activity occurs by embryonic day 11 soon after cessation of cell division, and an increase in high-affinity choline uptake and ACh synthesis and storage occurs just before hatching during the most active period of synaptogenesis. Of the amino acids studied, glycine, Gaba and β-alanine increase most markedly during retinal development. This finding is consistent with previous results suggesting a neurotransmitter role for Gaba, glycine and possibly β-alanine. In addition, a very large increase was observed in the retinal content of a substance tentatively identified as the dipeptide anserine. Further experiments will be required to determine whether anserine plays a role in retinal neurotransmission.

The amino acid developmental study was carried out with the technical assistance of Geoffrey Hastings. Support was provided by National Eye Institute Grants EY03502 and EY01995 and by the Sloan Foundation.

REFERENCES

Bader, C.R., R.W. Baughman and J.L. Moore. 1978. Different time course development for high-affinity choline uptake and choline acetyltransferase in the chick retina. Proc. Natl. Acad. Sci. USA 75: 2525-2529.

Baughman, R.W. and C.R. Bader. 1977. Biochemical characterization and cellular localization of the cholinergic system in the chicken retina. Brain Research 138; 469-485.

Baughman, R.W. and C.D. Gilbert. 1981. Aspartate and glutamate as possible neurotransmitters in the visual cortex. J. Neurosci. 1: 427-439.

Fisher, D.E., J.F. Amend and D.H. Strumeyer. 1977. Anserine and carnosine in chicks (Gallus gallus) rat pups (Rattus rattus) and ducklings (Anas platyrhynchos): comparative ontogenetic observations. Comp. Biochem. Physiol. 56B: 367-370.

Graham, L.T. 1974. Comparative aspects of neurotransmitters in the retina. In"The Eye", (H. Davson and L.T. Graham, Eds.), Vol. 6, Academic Press, New York, pp. 283-342.

Hebb, C. 1957. Biochemical evidence for the neural function of acetylcholine. Physiol. Rev. 37: 169-220.

Lindeman, V.F. 1947. The cholinesterse and acetylcholine content

of the chick retina, with special reference to functional
activity as indicated by the pupillary constrictor reflex.
Amer. J. Physiol. 148: 40-44.

Masland, R.H. and A. Ames III. 1976. Responses to acetylcholine
of ganglion cells in an isolated mammalian retina. J..
Neurophysiol. 39: 1220-1235.

Masland, R.H. and J.W. Mills. 1979. Autoradiographic
identification of acetylcholine in the rabbit retina. J. Cell
Biol. 83: 159-178.

Pasantes-Morales, H., J. Klethi, M. Ledig and P. Mandel. 1973.
Influence of light and dark on the free amino acid pattern of
the developing chick retina. Brain Research 57: 59-65.

Ross, C.D. and D.B. McDougal. 1976. The distribution of choline
acetyltransferase in vetebrate retina. J. Neurochem. 26:
521-526.

Control of Intercellular Communication via Gap Junctions

D.C. Spray, A.L. Harris, and M.V.L. Bennett

The gap junction is an extremely common structure which, it
is generally accepted, mediates intercellular communication (cf.
Bennett & Goodenough, 1977). While the function of the
intercellular exchange mediated by gap junctions may be very
different in various tissues, organisms and developmental stages,
many of the structural, biochemical and physiological properties
of gap junctions are generalizable across large phyletic distances
and from one tissue to another. Gap junctions are found
throughout the animal kingdom, from organisms as phylogentically
primitive as porifera (thin sections: Revel & Goodenough, 1970;
freeze fracture, Ginzberg & Morales, unpublished) to those as
advanced as arthropods, molluscs, and higher vertebrates (cf.
Staehelin, 1974). At least some cells in virtually all vertebrate
organ systems are linked by gap junctions. Notable groups of
cells that are not coupled include striated muscle fibers,
differentiating spermatozoa, circulating erythrocytes and specific
neurons (cf. Gilula, 1978; Revel, 1978). The near ubiquity of the
gap junction and its evolutionary conservation imply a fundamental
importance of this structure to cellular function.

   Structural information about the gap junction has been
obtained from freeze fracture and thin section images, both
positively and negatively stained (Figure 1). With the freeze
fracture technique, gap junctions are usually identifiable as
discoid arrays of large (8-10 nm) intermembrane particles (Figure
1A). These particles are attached to the E face in anthropods,
but in most other organisms, including all vertebrates, the
intramembrane particles cleave with the P face, leaving
corresponding pits in the E face. Often a central depression is
visualized in the center of each particle. In thin section the

gap junction is defined as a generally linear close apposition of the membranes of two cells to within 2 to 4 nm (Figure 1B). The resulting image is septilaminar, with the width of the "gap" between the unit membranes about the same size or slightly less than the width of the unstained membrane interior. When tissues are perfused with the electron dense extracellular space markers lanthanum or ruthenium red prior to fixation, a somewhat periodic substructure that bridges the gap is discernable (cf. Revel & Karnovsky, 1967). Negatively stained images of isolated junctional membrane reveal hexagonally packed toroids with electron dense centers and outlines (cf. Hertzberg & Gilula, 1979). On the basis of Fourier analysis of images of negatively stained junctions (Unwin & Zampighi, 1980) and x-ray diffraction patterns (Makowski et al., 1977) each toroid is believed to possess radial six-fold symmetry.

The particles seen in freeze fracture, the periodic substructure of the gap seen against a lanthanum background and the toroid shapes seen in negatively stained isolated junctions are believed to represent different images of the same structure. The emergent model (Figure 1C) has the following characteristics: Each membrane contains a hemichannel with a hexameric arrangement of identical subunits, each of which is asymmetric with regard to cytoplasmic and extracellular surfaces. Series connection of two apposing hemichannels at their extracellular faces is thought to form a single gap junction channel or connexon. If each connexon does represent a single channel, cells could be weakly coupled by a single particle and the degree of coupling would then increase with increasing number of connexons. For cells with high input resistance, coupling sufficient to synchronize electrical activity might require only a few functional connexons and the structural correlate might be hard to find (cf. Williams & DeHaan, 1981).

Figure 1. Structure of the gap junction. A. Freeze fracture replica of a gap junction from a Fundulus cleavage stage embryo showing P-face particles (top) and E-face pits. Micrograph and preparation by R.D. Ginzberg. B. Thin section of large gap junction between axolotl blastomeres (Harrison stages 6 to 8); magnification as in A. (From Hanna et al., 1980.) C. Schematic diagram illustrating one model of the arrangement of the gap junction channel in the membrane. Each membrane contributes half the channel, and each of these hemichannels is comprised of six identical subunits surrounding the aqueous channel interior. (From Makowski et al., 1977.)

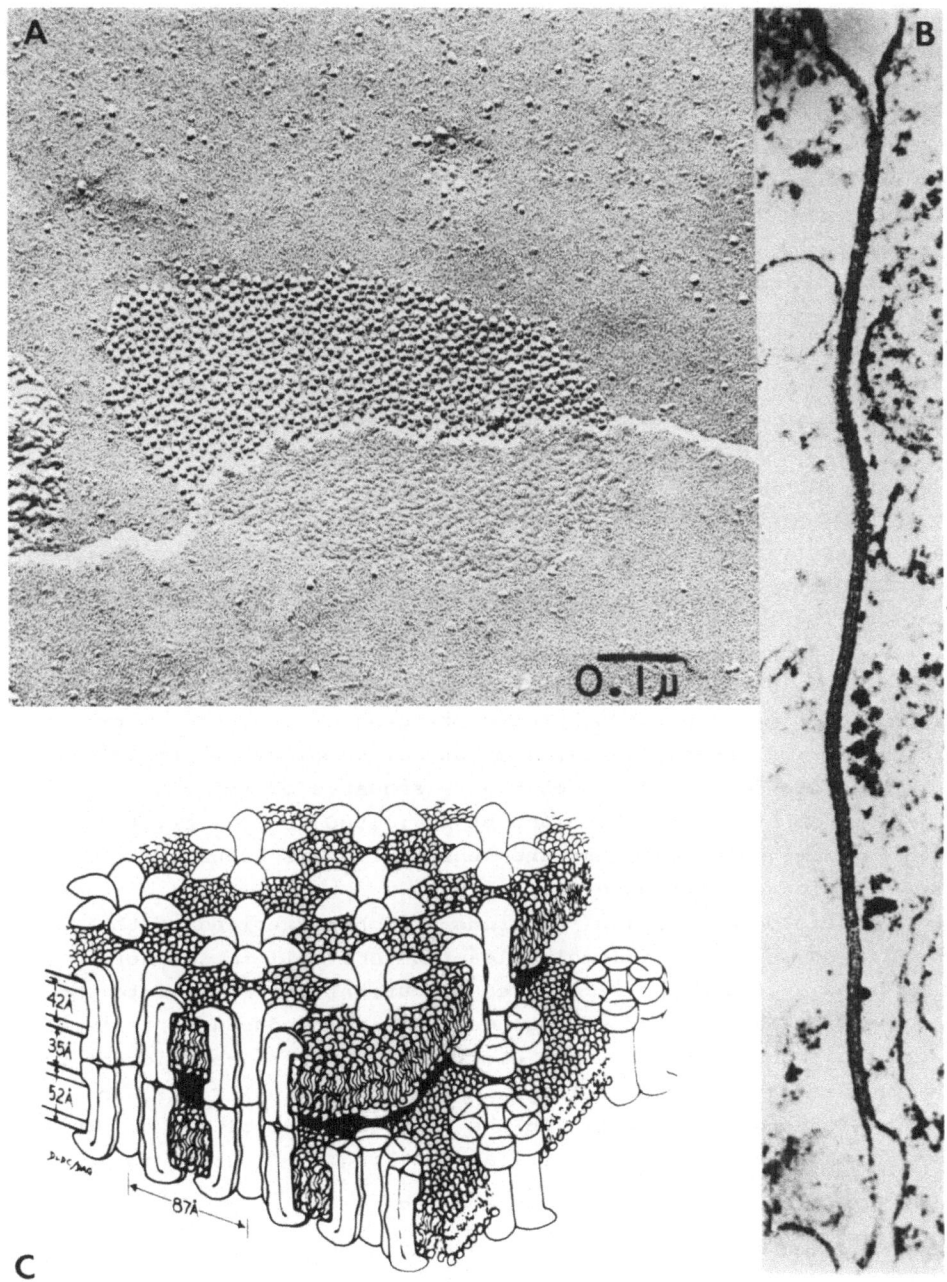

Biochemically, the major proteins present in isolated lens and liver gap junctions have been characterized with regard to molecular weight. The major integral membrane protein in liver gap junctions has an apparent molecular weight of about 27 kD whereas that from lens is somewhat smaller (25-26 kD; Finbow et al., 1980; Hertzberg & Gilula, 1979). (Earlier reports of smaller major junctional proteins probably can now be discounted as artefactual results of proteolytic digestion during isolation procedures.) Glycosylation of the major junctional protein is probably minor (Herzberg & Gilula, 1979) and the in situ lipid environment is apparently unremarkable (Revel et al., 1978). It is therefore concluded that the gap junctional proteins isolated from liver and lens are primarily polypeptides composed of as many as 300 amino acid residues. In the gap junction model of Figure 1C, each of these polypeptides could represent a monomer, the hexamer of which comprises the hemichannel, so that the total molecular weight of the hypothesized channel would be about 300kD (or about twice as large as the ACh channel isolated from Torpedo electroplaque, a molecule which spans a single membrane (cf. Karlin, 1980). Although similar in size, other biochemical evidence suggests that proteins isolated from liver and lens are different, including a consistent molecular weight difference of 5-12%, dissimilar electrophoretic patterns of products of protease digestion (Hertzberg, 1980), and lack of homology in the 20% of the molecule for which a preliminary sequence of amino acid residues is available (Nicolson et al., 1980). However, there is not complete agreement that the lens protein is actually that of a gap junction (Robertson et al., 1981).

Physiologically, gap junctions are characterized by their ability to pass ions and molecules from one cell to another. Electrically, coupling is assayed by current injection into one cell and measurement of voltage deflections produced in that cell and those coupled to it. The electrical coupling coefficient is defined as the ratio of voltage in the coupled cells (V2) to that of the injected cell (V_1). In the simplest case of two coupled isopotential cells, the electrical coupling coefficient is

$$k_{21} = V_2/V_1 = g_j/(g_j + g_2)$$

where g_j and g_2 are ionic conductances of the junctional membrane and the nonjunctional membrane of the second cell, respectively (Figure 2D). When measurements are taken at a time that is long

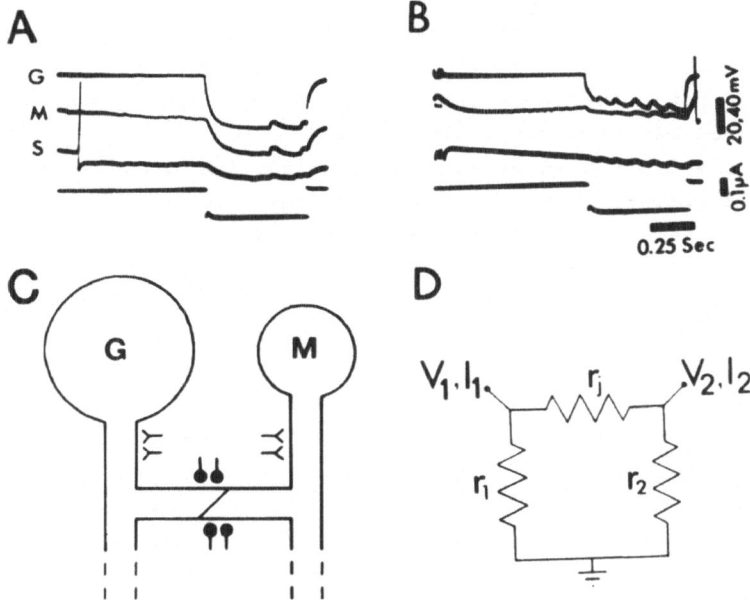

Figure 2. Uncoupling of neurons in Navanax by chemical synaptic input (modified from Spira et al., 1980). A,B. Recordings from ipsilateral expansion motoneurons appear on the first three traces, and the 0.5 sec pulse of current injected into the G cell is shown on the lowest trace. A. Coupling in the resting condition. B. Following a brief train of stimuli to the pharyngeal nerve the input resistance of the G cell was reduced and current no longer spread to M and S cells, which were uncoupled from G. C. Diagram of motoneurons coupled by axon collaterals showing the hypothesized location of uncoupling synapses. Inhibitory synapses (filled circles) near the site of coupling (diagonal line) can shunt the intercellular spread of current without blocking excitation by synaptic input (Ys) near the cell body. D. The simplified electrical circuit represents the nonjunctional (r_1,r_2) and junctional resistance (r_j) of two coupled isopotential cells. While this program illustrates the role that nonjunctional conductance can play in determining the strength of coupling, a more realistic circuit has been presented elsewhere which allows cells to be excited while uncoupled (Spira et al., 1980).

compared to the membrane time constant, as is the usual practice, nonjunctional capacitance can be ignored. From this equation it is obvious that cells can be electrically uncoupled by treatments that either decrease the conductance of the gap junctional membrane or increase the conductance of nonjunctional membrane. As discussed below, changes in non-junctional conductance can control transfer of information from one nerve cell to another by short circuiting current in the coupling pathway (Spira et al., 1980). Also, gross damage to the nonjunctional membrane of a cell can eliminate coupling without changing conductance of the junctional membrane itself.

The extent of intercellular coupling may also be assayed by the intercellular spread of dyes or other compounds. For a pair of cells where one cell generates or is injected with a compound, and the concentration within each cell is equalized by diffusion, the chemical coupling coefficient is defined as the ratio of the steady state concentration of that compound in the recipient cell to that in the source cell. Like electrical coupling, strength of chemical coupling depends upon relative permeabilities of junctional and nonjunctional membranes to the indicator compound, but the chemical coupling coefficient can also be altered by cytoplasmic binding or metabolic degradation. Another essential difference between these two indices of coupling is dependence of the detection of chemical coupling on the relative volumes and specific geometries of the donor and recipient cell; the time constants for dye spread therefore can be many minutes or even hours compared with a much shorter time constant of electrotonic spread (milliseconds). The dependence of electrical and chemical coupling coefficients on different parameters can lead to cases in which they differ. If a junction-permeant dye (such as Lucifer yellow CH) which has no appreciable binding, self-quenching, cellular degradation, or nonjunctional permeability (cf. Stewart, 1981) is injected briefly into one cell of a pair of even poorly coupled cells, its concentration in the two cells must eventually equilibrate. The final result of dye equilibration is independent of the strength of electrical coupling. Alternatively, if cells are closely coupled electrically they may not be chemically coupled if the dye used either readily crosses the nonjunctional membrane or fails to cross the junctional membrane. Cells may also appear not to be coupled if insufficient transfer time is allowed (Bennett et al., 1978a). These experimental difficulties may account for observations that dye coupling may be totally lost under conditions in which electrical coupling is only somewhat reduced (Rose & Loewenstein, 1976; Goodman & Spitzer, 1979). While these results were interpreted to indicate partial closure of channels, our own prejudice is that closure of the gap junctional channel is an all-or-none event, as is true of channels

in other biological tissues (cf. Neher and Stevens, 1977).
Resolution of this issue will require correlation of flux rate
with junctional conductance.

From physiological data, the gap junctional channel seems to
be an aqueous pore providing a low resistance diffusional pathway
from one cell to another without access to extracellular space.
The limiting internal diameter of this channel is less than 1.4
nm, so that spherical molecules up to a molecular weight of about
1 kD are permeable. Current-carrying ions such as Na^+, K^+, and
Cl^- are well within this size range (cf. Bennett, 1977), as are
mono- and disaccharides, nucleotides, cyclic AMP, various high
fluorescent dyes such as Lucifer yellow CH, and short polypeptides
chains (cf. Simpson et al., 1977; Bennett, 1977). Gap junctions
are not permeable to nucleic acids, large polypeptides, or
cellular organelles (Pitts & Simms, 1977).

CONTROLS ON COUPLING

Although there is considerable debate over the role that gap
junctions play in cellular function, that role presumably involves
the exchange of information contained in the moderately sized
molecules that can diffuse (or conceivably be iontophoresed) from
one cell to another. Controls on intercellular communication can
be dichotomized as (1) those which change the number of channels
in the membrane (formation and loss of junctional material), and
(2) those which influence the coupling either by changing
conductance of nonjunctional or junctional membranes.

Formation and Loss

Formation and loss of gap junctions between cells are well
recognized cellular processes and may occur under hormonal or
experimental control. A striking illustration is the very rapid
turnover time of liver junctions whose half life is about 5 hours
(Fallon & Goodenough, 1981). A number of issues remain
unresolved. Formation presumably begins with the synthesis of
junctional monomers on ribosomes within the cell; it remains to be
determined where and when monomers are inserted into the membrane
and organized into hexamers. In some tissues, the first
recognizable step in junctional formation is the appearance of
small gap junctions (cf. Ne'eman et al., 1980). In other tissues
larger precursor particles appear (Johnson et al., 1974; Decker,

1976). The regions where the larger (11 nm) particles are found have been termed "formation plaques" and are common preliminary stages in regenerating liver and between Novikoff hepatoma cells (Yee & Revel, 1978; Johnson et al., 1974). In most tissues studied, junction formation proceeds with the accretion of additional particles or fusion of small plaques of normal size.

Electronic coupling can be present within a few minutes following reassociation of single Fundulus blastomeres (e.g. Ne'eman et al., 1980). De novo protein synthesis is not required for the rapidly forming junctions of the hepatoma cell line (Epstein et al., 1977) but is essential for thyroxine-induced formation of junctions between tadpole ependymoglial cells (Decker, 1976) and for acquisition of beat synchrony in cultured heart cells (Griepp & Bernfield, 1978). Whether these different sensitivities to protein synthesis inhibitors reflect a real difference in de novo gap junction production, necessary synthesis of proteins involved in cell contact in some systems, or differences in capacities of precursor pools is still conjectural.

In some tissues, the rate of gap junction formation can be modulated by specific treatments, among the most dramatic of which occurs in the compensatory regeneration after partial hepatectomy. Removal of the anterior and medial lobes of the liver in rats initiates, through an as yet unknown stimulus, a profound growth of the remaining tissue. There is an initial loss of junctions followed within 48 hours by gap junction reappearance (Yee & Revel, 1978). Whether the initiating signal is local or generated elsewhere is still unclear. Gap junctions in some other tissues prolifererate following treatment with hormones. Granulosa cells in ovaries of hypophysectomized rats proliferate and increase their gap junctions in response to gonadotropin and estrogen (Merk & McNutt, 1972), dye spread between oocyte and cumulus cells in Xenopus ovary is increased by gonadotropin (Browne et al., 1979), and in myometrium of rat uterus, estrogens (perhaps acting through prostaglandins) increase and progesterone inhibits gap junction formation suggesting a role in parturition (Garfield et al., 1980). Pancreatic β-cell gap junctions are increased in number and area in response to insulin (Meda et al., 1979). Furthermore, vitamin A induces metaplasia and consequent gap junction formation between cultured epidermal cells (Elias & Friend, 1976). Preliminary electrophysiological evidence has appeared for opposite effects of cyclic AMP and cyclic GMP on gap junction

formation among Novikoff cells. Dibutyryl cAMP (1mM) increases
the rate of coupling whereas 8-bromo cGMP (less than 1μM)
decreases formation (cf. Sheridan, 1978). In reaggregates of
isolated heart cells, trypsin accelerates the entrainment of
synchronous beating whereas trypsin inhibitors block the synchrony
(Griepp & Revel, 1977). While these data may suggest involvement
of an enzymatic step in junctional formation, other possible sites
of trypsin action include removal of surface material that may
sterically hinder cell contact.

The processes by which gap junctions are lost appear to be
several. Internalized gap junctions are seen in cytoplasmic
vesicles in granulosa cells during oocyte maturation (Albertini &
Anderson, 1975) and in developing ootocyst epithelium (Ginzberg &
Gilula, 1979), implying endocytosis and perhaps eventual lysosomal
degradation. Sealed cellular remnants that interface with intact
cells by gap junctions are seen on the surfaces of pancreatic
cells in a few minutes after mechanical dissociation (Amsterdam &
Jamieson, 1974). In some other systems the initial event in
junctional disappearance involves the splitting of the connexon
into hemichannels (Lane & Swales, 1978; Goodenough & Gilula, 1974;
Hanna et al., 1979). In these systems the particles may
secondarily disperse, suggesting that the junctional material
might remain in the membrane. The basic questions remain whether
the cytoplasmic and membrane precursor pools exist and are
exchangable and reusable and whether equilibrum between these
pools is affected by specific stimuli thus far employed in the
study of junctional formation and breakdown.

Uncoupling by Increased Nonjunctional Conductance

A role played by gap junctions between excitable cells is the
spread of currents along a conduction pathway or among a
population of cells. While communication by way of chemical
synapses could in many cases replace electrotonic interactions,
the speed and synchronizing effects are apparent evolutionary
advantages for electrotonic interaction. Since spread of current
among coupled cells depends on both junctional and nonjunctional
conductances, chemical synaptic inputs might shunt current flow
(Fig. 2).

This shunting effect of chemical synaptic inputs apparently
plays a role in the feeding behavior of the mollusc, Navanax

(Spira & Bennett, 1972; Spira et al., 1980). The feeding behavior
is initiated by sufficient contact of cephalic sensory areas with
prey to excite an electrically coupled pool of motoneurons
innervating radially arranged muscles forming most of the bulk of
the muscular pharyngeal tube. Excitation of these motoneurons
contracts the radial musculature, rapidly expanding the pharynx
and sucking prey inside. Mechanoreceptors with fields in the
pharyngeal wall are activated by ingestion (Spira et al., 1980)
and feed back inhibitory chemical synaptic input in the region of
the gap junctions. This increased conductance to current-carrying
ions shunts current in this region, so that the cells are
uncoupled (Fig. 2; Spira et al., 1980). This uncoupling
apparently allows the neurons to fire asynchronously, permitting
peristalsis. In the mammalian inferior olive, uncoupling by
strategically placed inhibitory inputs was inferred on an
anatomical basis: synapses with appropriate morphology occur near
dendrodendritic gap junctions (Sotelo et al., 1974). Similarly,
rectification of nonjunctional conductance may account for the
phenomenon of rectification in some synapses in leech (Zipser,
1979), current of one polarity evoking a conductance increase and
thereby shunting itself. Activation of conductance decreasing
synapses would be expected to increase coupling; such a situation
has been found in neurons controlling inking behavior in Aplysia
(Carew and Kandel, 1976).

These changes in coupling by changes in nonjunctional
conductance are of interest because gap junctions in the nervous
system appear to mediate rapid spread of current from one cell to
another. Uncoupling of pancreatic acinar cells by acetylcholine
may be a result of increase in nonjunctional conductance (Iwatsuki
& Petersen, 1976). In this system the electrical uncoupling may
be an epiphenomenon since it would presumably not affect metabolic
coupling mediated by the gap junctions. Studies that assay
effects of pharmacological compounds on junctional conductance by
changes in the coupling coefficient should thus be interpreted
with the caution that nonjunction conductance must remain constant
in order for the measure to be valid.

Uncoupling by Decreased Junctional Conductance

Junctional conductance can be reduced by a variety of
treatments, such as increasing intracellular concentrations of

hydrogen or calcium ions, exposure to various pharmacological
agents, and, in certain systems, by applying a voltage across the
junction.

pH Dependence of Junctional Conductance

The coupling between cells in a number of systems is reduced
by lowering the pH of the cytoplasm (pH_i) (cf. Spray et al.,
1982b). We have used pairs of large blastomeres dissected from
early teleost or amphibian embryos with which it was possible to
accurately measure junctional conductance either under voltage
clamp (Spray et al., 1981a) or with current injection accompanied
by solution of the simple equivalent circuit (cf. Bennett, 1966).
In many of these experiments pH_i of one of the cells was measured
with a Thomas-type pH microelectrode.

Most cells are quite impermeable to H ions but are permeable
to the un-ionized forms of weak acids. A simple method of
decreasing pH_i is, therefore, to bathe the cells in a weak acid at
a pH at which some proportion of it is un-ionized. CO_2 and the
un-ionized forms of acetate, lactate and propionate, for example,
are capable of diffusing across the cell membrane where they enter
into a new equilibrium and release H ions. Another method is to
directly inject H ions into a cell. External H ions have little
effect on junctional conductance as shown by bathing the cells
with low pH solutions containing strong (and thus almost
completely ionized) acids or impermeant buffers.

Junctional conductance is a simple function of pH_i; the
conductance drops as pH_i drops and recovers as pH_i recovers
without hysteresis (Figure 3A). The changes can be produced by
any of the weak acids tested and the efficacy of the weak acid is
inversely related to its degree of ionization in the external
medium. The pH_i-conductance relation is similar when obtained
under voltage clamp conditions (Spray et al., 1982a),
demonstrating a lack of dependence on the depolarization that
often accompanies application of weak acids. The relation between
pH_i and junctional conductance is fit well by a Hill plot with an
apparent pK of 7.3 (the pH at half-maximal conductance) and Hill
coefficient of 4 to 5 (Fig. 3A). The results are consistent with
a model in which H ions interact with four or five highly
cooperative sites in closing the junctional channels, but there
might be more sites operating with less cooperativity.

Intracellular injection of H ions also decreased junctional conductance in Fundulus blastomeres. This change is relatively stable, also reflecting H^+ impermeability of the membranes, but is rapidly reversed when pH_i is raised by bathing with ammonium salts (the un-ionized NH_3 permeates the surface membrane and neutralizes internal H ions in a manner analogous to acidification by weak acids).

The rapid action of changes in pH_i and its simple relation to conductance implies a direct effect on the junctions rather than mediation by a cytoplasmic intermediate. We have recently developed a preparation in which one cytoplasmic aspect of the junctions (in Fundulus) is directly perfused to wash away most or all of the cytoplasm (Stern et al., 1980; Spray et al., 1982b). With Ca^{++} held constant at 1 μM, low pH solutions have the same effect on the junctions in this preparation as in intact cells which makes the hypothesis of a cytoplasmic intermediate even more unlikely (Fig. 3B, inset).

Figure 3. Sensitivity of junctional conductance to intracellular H and Ca ions. A. In one of a pair of coupled cells pH_i was measured with an intracellular electrode during perfusion with CO_2 containing saline and subsequent recovery in normal saline. CO_2 uncouples the cells. With separate voltage and current electrodes in each cell the voltage responses to alternate current pulses were used to calculate junctional conductance, which is plotted as a function of pH_i. Triangles with apices down represent initial values and values during cytoplasmic acidification. Triangles with apices up are values during recovery. The points fall along Hill curves with $pK_H=7.3$ and n=4 and 5. There is no evidence of hysteresis between falling and rising pH_i. The data are consistent with H ions acting directly on a channel macromolecule. (From Spray et al., 1981a.) B. From cell pairs in which one cytoplasmic aspect of the junctions was perfused with a solution buffered to pH 7.8 and various pCa, starting at higher pCa, lowering pCa and then returning. Separate series for pCa 4.0 to 6.0 and 3.0 and 3.9. The ordinate g_p/g_n is a normalized measure of junctional conductance assumed equal for pCa 3.9 and 4.0. Standard errors and number of experiments are indicated. The curves are Hill plots fit by eye with the assumption that the minimum value of g_p/g_n was either zero or just below that at pCa 3.0. Junctional conductance is very insensitive to Ca as compared to H ions (Inset) Experiments similar to those in B except that pCa was held at 7.0 while pH was changed between 7.8 and 6.8. The ordinate G again represents a normalized value of conductance. The smooth curve is the Hill plot for $pK_H=7.3$ and n=4.5 obtained for intact cell pairs (see A). The junctional conductance at pH 7.2 (error bar for 6 trials) falls close to the curve indicating that the perfused preparation has the same pH sensitivity. (From Spray et al., 1982b.)

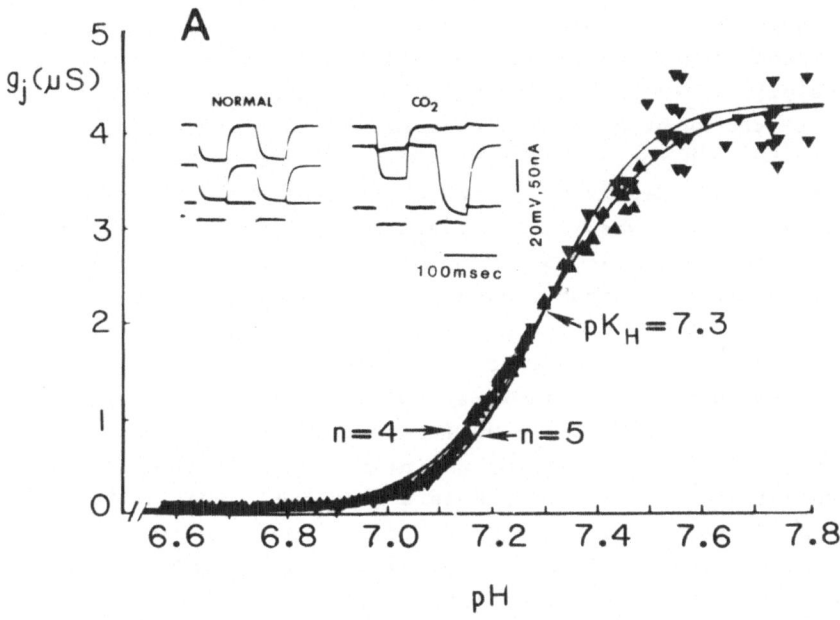

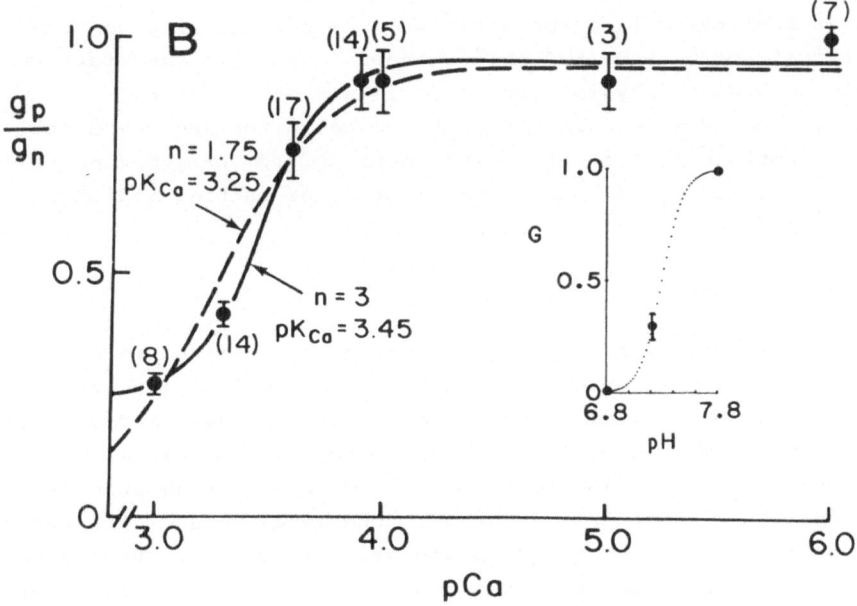

Ca Dependence

Injection of Ca ions is also known to decrease the
conductance of gap junctions (De Mello, 1975; Rose and
Loewenstein, 1976), and since buffering systems for H and Ca
interact, the question arises whether each acts independently or
one acts through changing the other. With the perfused
preparation we independently varied the free Ca^{++} and free H
activities. There was little effect on the junctions of Ca^{++}
levels up to 0.1 mM, and some conductance apparently remained at 1
mM Ca (the highest concentration we used, Fig. 3B). These data
indicate that gap junctions are about ten thousand times less
sensitive to Ca than to H ions, and make it very unlikely that H
ions act through Ca^{++} as an intermediate. Intact cells show a
similar sensitivity to Ca/Ca-buffer injections. Direct
measurements of Ca^{++} levels by an intercellular ion-sensitive
electrode (Rink et al., 1980; Hess & Weingart, 1980) or by
aequorin (Bennett et al., 1978) failed to reveal rises in Ca^{++}
during acidification of the cytoplasm that decreased junctional
conductance. The high concentrations of Ca^{++} that are required in
Fundulus would certainly have been observed.

The relative insensitivity of the junctions to Ca ions
implies that rises in cytoplasmic Ca^{++} do not normally control
junctional conductance and that changes in pH_i are therefore more
likely to mediate physiological regulation of junctional
conductance. However, in the pathological situation where one
cell's membrane is disrupted, its coupling from neighboring cells
may be mediated by Ca ions entering from the bathing medium
without substantial change in pH_i.

Voltage Dependence

Another way that coupling is controlled is found in amphibian
embryos in which junctional conductance is highly sensitive to
transjunctional voltage (Spray et al., 1979; 1981a; Harris et al.,
1981). A new method of dual voltage clamp allowed us to
characterize the voltage sensitivity. When a voltage step is
applied to the junctional membrane, junctional conductance relaxes
exponentially to a new steady state value (Fig. 4A). From steady
state values (Fig. 4B) and time constants of change, opening and
closing rates were calculated and found to be exponential
functions of voltage. The data could be explained in terms of a

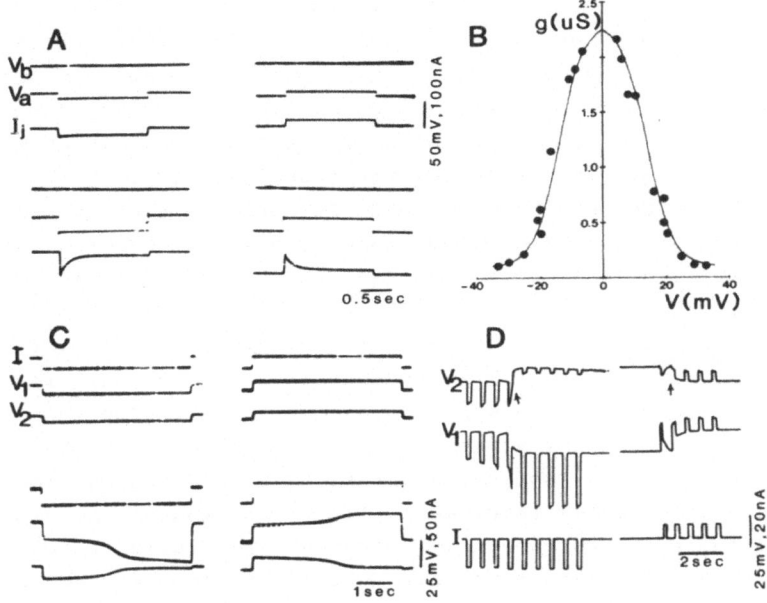

Figure 4. Voltage dependence of junctional conductance in amphibian blastomeres. A. Records from a dual voltage clamp experiment on a coupled cell pair. One cell was stepped to a new voltage, while the second cell was held constant. The current applied in the second cell represents the current flowing through the junction as a result of the voltage step. For small voltages of either sign the current decreases only slightly during the step. For larger voltages of either sign the current relaxes exponentially to a lower value. B. Steady state junctional conductance g_j as a function of transjunctional voltage. g_j decreases steeply for voltages of either sign. The curve is a Boltzmann relation as would be observed if the energy difference between open and closed stages were a linear function of transjunctional voltage. The parameters of the particular curve are for the equivalent of about 6 electron charges moving through the entire transjunctional voltage. (From Spray et al., 1981b.) C. Effects of current pulse. During large enough current pulses of either sign, the voltage in the directly polarized cell increases in a regenerative manner while the voltage in the other cell declines. The cells change from a well coupled to a poorly coupled state as a result of voltage dependence of the junctional conductance. (From Spray et al., 1979.) D. Bistability of uncoupling resulting from voltage dependence and difference in resting potentials between a pair of cells. A train of pulses in one cell (V_2) leads to progressive decrease in junctional conductance until the cells uncouple, the pulsed cell having the more negative resting potential. The break in the record represents 100 sec. Near the end of the record an oppositely directed pulse was applied that reduced the transjunctional voltage and allowed the cells to recouple.

channel macromolecule that changed its dipole moment depending on
its conductance state. The effects of reversal of transjunctional
voltage are consistent with there being two gates in series in
each channel that sense local voltage within the aqueous pore.
This model is also consistent with the symmetry of gap junctions,
in which each membrane contains half the junctional channel.

The voltage dependence may function in controlling
communication between blastomeres. As would be expected, dye
molecules that cross high conductance junctions do not cross them
when they are held at low conductance by transjunctional voltage
(Spray et al., 1979).

In current clamp conditions the voltage dependence of
junctional conductance allows regenerative uncoupling. In the
cell in which current is applied responses to long current steps
have the appearance of action potentials while in the other cell
the electrotonically spread potential falls off as junctional
conductance decreases (Fig. 4C). If the resting potentials of
coupled cells are sufficiently different, the cells exhibit
bistability and remain in either coupled or uncoupled conditions
without application of extrinsic current (Fig. 4D). Since
substantial resting potential differences can occur during
amphibian development (Warner, 1973), voltage controlled
uncoupling might be involved in separating populations of cells.
There are no data as to whether the voltage dependence is found in
adult amphibians, although it is found in Rohon-Beard neurons in
Xenopus 24 hours after fertilization (Spitzer, 1980).

Multiplicity of Gating Mechanisms

Although there is as yet little direct evidence, our
hypothesis is that junctional channels fluctuate between open and
closed states. The response to increased cytoplasmic H and Ca
ions or transjunctional voltage would therefore be a shift in the
distribution toward the closed state. It is reasonable that H and
Ca ions act on the same sites since competition between these ions
is a commonly observed property. The slope of the
conductance-concentration relationship at the K_m for Ca^{++},
although not yet well defined, is about half as great as that at
the K_m for H ions, so that two H ions or one Ca ion could be
acting on a common site, Ca of course having a much lower
affinity.

A number of data indicate that the voltage sensitivity in amphibian cells is mediated by a separate gating mechanism from that operating in pH dependence. As conductance is reduced by lowering pH_i, the remaining conductance exhibits the same voltage sensitivity as did the maximum conductance; this behavior would be observed if low pH_i simply closed a gate in series with the voltage sensitive gate (Spray et al., 1981c). Some of the treatments described in the next section also act independently on pH and voltage sensitivity and support the concept of separate gating mechanisms.

Chemical and Pharmacological Treatments That Decrease Junctional Conductance

One attractive aspect of the controls on junctional conductance described above is that pharmacological treatments can be applied and their effects determined on sensitivity to voltage and pH as well as simply on conductance. A class of chemical treatments that earlier was shown to decrease junctional conductance is fixation by aldehydes or other agents used in tissue preparation for electronmicroscopy (Bennett et al., 1972). An important implication of these findings is that fixed junctions are of low conductance, i.e., their channels are closed before they are frozen or dehydrated, and any structural correlates of junctional state must be considered in this light. The new technique of rapid freezing allows one to bypass the changes that occur with chemical fixation (Heuser et al., 1975, see below). Although the reaction of aldehydes with amine groups releases protons, the decrease in conductance is not due to decreased cytoplasmic pH as determined by direct measurement (Spray et al., 1981c).

Lower concentrations of aldehydes give interestingly graded effects. Glutaraldehyde applied externally at 10-20 µM for several minutes irreversibly reduces junctional conductance between fish or amphibian blastomeres by 50-80%. In both tissues the remaining conductance is not due to cytoplasmic continuity resulting from cell fusion at the membrane appositions (Bennett, 1973b), since it is blocked by fixative concentrations of glutaraldehyde. Furthermore, in amphibian blastomeres voltage sensitivity is unaffected in respect to both steady state values and rates (Spray et al., 1981c).

The effects of formaldehyde are distinctively different. Comparable reductions in conductance after comparable exposure times require concentrations near 10mM. The remaining conductance is as sensitive to CO_2 solutions as before exposure to aldehyde. In contrast the voltage dependent changes in amphibian cells are markedly slowed (Spray et al., 1981d). Quantitative analysis of steady state conductances, time constants, and opening and closing rates are in progress. The specific actions of the different aldehydes on pH and voltage sensitivities suggest that the aldehydes act at different sites on the junctional macromolecules.

A number of anesthetic agents are effective in reducing junctional conductance. Octanol reduces junctional conductance by half at concentrations of about 50 μM in blastomeres (Spray et al., 1981d) and 1mM in crayfish septate axon (Ramon et al., 1980). Heptanol and lower alcohols are much less effective in crayfish. In blastomeres tricaine at 10-100. μM and benzocaine at 2-5mM also reduce junctional conductance whereas procaine, tetracaine, and lidocaine (at 10mM) do not. None of the local anesthetics used had clear effects on voltage dependence of junctional conductance or sensitivity to CO_2 solutions. These compounds may act by a different mechanism than H and Ca ions or transjunctional voltage.

Structural Correlates of Gating

In blastomeres, decrease in junctional conductance by cytoplasmic acidification is not associated with obvious changes in structure (Hanna et al., 1978), and recovery of conductance on restoring pH_i is much more rapid than formation of new junctions. Changes in particle structure with conductance state are minimal, even with the more refined techniques of fast freezing and low temperature circular shadowing (Hanna et al., 1981). In some tissues similar treatment applied over longer periods produces regularization of the particle arrays, which appears to be a slow process secondary to the conductance changes (Raviola et al., 1980, see Peracchia, 1980). It is not clear that the regularization is reversible; instead it may provide a signal for internalization of junctional material. Many more studies correlating morphological and physiological data will be required to resolve these issues.

GAP JUNCTIONS AND DEVELOPMENT, POSSIBILITY FOR CONTROL

The permeability of gap junctions to moderately large
molecules and the existence of treatments that control junctional
conductance confer a considerable capacity for signalling on the
gap junction channel. A pivotal role for communication via gap
junctions in differentiation and development is suggested by their
widespread occurrence in embryos (Furshpan & Potter, 1968;
Bennett, 1973a; Lo & Gilula, 1980). Furthermore, the role of gap
junctions in metabolic cooperation in tissue culture is well
established (Gilula et al., 1972; Pitts & Simms, 1977). They can
transmit second messengers of hormonal stimuli (Tsien & Weingart,
1976; Lawrence et al., 1978), and in a number of cases the
junctions appear transiently at times when it would be reasonable
to suppose that a message was being transferred (cf. Bennett et
al., 1981). While the properties of gap junctions seem ideal for
the control necessary during development, the evidence that gap
junctions actually play such a communicative role remains
correlative.

There are various ways in which chemical communication
through gap junctions might have relevance for development. The
brief passage of a qualitative message or signal from one cell to
another to trigger a specific behavior of the contacted cell is
suggested by studies on the behaviors of individual neurons in
development. Electron microscopic studies show that as the eight
optic nerve fibers from a single Daphnia ommatidum enter the optic
lobe, one becomes the lead fiber and transiently forms gap
junctions with precisely five neuroblasts (LoPresti et al., 1974).
These five cells become a "cartridge" receiving axons from the
same ommatidum; the lead fiber subsequently synapses with only one
of them. The lead fiber triggers axonal growth in the neuroblasts
as it successively contacts them and may transmit a molecular
message that determines that they form appropriate synapses. The
neuroblasts may also give a message back to the lead fiber that
determines its behavior, for the lead fiber in some way counts the
number of target neuroblasts that it contacts. Similar examples
of specific and transient coupling occur during differentiation of
the grasshopper nerve cord (Raper & Goodman, this volume).

A molecular message crossing gap junctions may operate in
cell recognition. In tissue culture, nerve cells occasionally
couple to muscle cells on which they also form chemical synapses

(Fishbach et al., 1974). A message from the muscle fibers may tell the nerve tthat it has found an appropriate place to form a synapse. Nerves form close appositions with muscle fibers at an early stage of innvervation in the embryo (Kelly & Zacks, 1969), but gap junctions have not yet been demonstrated clearly. A counterexample occurs in the chick ciliary ganglion in which electrical coupling develops only after chemical synapses are well established (Landmesser & Pilar, 1972), but an early transient stage of weak electrical coupling (that could still produce a strong chemically mediated effect) cannot be excluded.

Another form of chemical communication via gap junctions may be the establishment of a field. A cell may be in a chemical gradient established by diffusion through gap junctions from a source region and the cells' response may be a function of the concentration. Such a gradient could provide positional information, which appears essential to development (Wolpert, 1978). A primary example for this form of interaction has been shown in the amphibian retina. Gap junctions disappear in the central retina shortly after what had been presumed to be specification of the axes, based on physiological data, becomes irreversible (Hunt & Jacobson, 1974). Recently this interpretation has been questioned because respecification of the axes may be an artifact of regeneration from unrotated tissue (Cook, 1980; Sharma, this volume). Prior to neurulation in Xenopus, coupling is essentially uniform across lateral and presumptive neural ectoderm, and a gradient in resting membrane potential is observed which correlates with the presumptive fates of the epithelium (Warner, 1973).

Regeneration in hydra has been analyzed in terms of positional information (Wolpert, 1974), and it is one of the new animals from which candidate morphogens have been isolated (Grimmelikhuijzen and Schaller, 1979). There are four of these substances with molecular weights between 500 and 1000 D, which are within the size range expected to cross gap junctions (see size limit of permeable molecules above) which connect the epithelial cells (Wood, 1979). Even though gap junctions could presumably pass the morphogens in hydra, there is still no very good reason to believe that they do so. The morphogens are normally concentrated in the neurons, which probably are not connected to the epithelial cells by gap junctions, although in nerve free hydra the morphogens persist and presumably are made by epithelial cells. The assays for morphogenetic actions involve surface application, more consistent with action across extracellular space as in neurosecretion than with intercytoplasmic diffusion. Also, the space between endo-and ectoderm forms a restricted pathway which might be well suited to establishment of morphogen gradients.

Spatial concentration patterns in an essentially uniform field may be established through chemical reactions involving inhibitory and activator substances (Gierer, 1977). If the activator is generated autocatalytically while the inhibitor is not, the inhibitor is capable of spreading further. One can obtain a variety of patterns that depend not only on chemical coupling coefficients, but on the extent of the field. Thus, if chemical coupling were mediated by gap junctions, the nature of the pattern would depend on the extent and strength of coupling. Modulation of coupling by any of the physiological controls described above could then have profound effects.

In respect to control of coupling it may be relevant that neurulation in amphibia is associated with the generation of large resting potentials by turn on of sodium pumps in the infolding cells (cf. Warner, 1979). Although the role of sodium pump activation was interpreted to be in changing concentrations of intercellular ions, the observed changes in membrane potential would be sufficient to clase gap junction channels (Spray et al., 1981b). While intracellular pH is known to decrease from early embryo to differentiated adult, regional pH_i changes during development have not been determined. In such changes may also lie a cause of modulation of couping and a stimulus for differentiation.

Several qualifications should be made concerning the evidence for a role of gap junctions in development. Although chemical communication via gap junctions is virtually unknown except in artificial circumstances, we do know of chemical communications across extracellular space, as at chemical synapses or by hormones. Differentiation of cells, such as sympathetic neurons, can be influenced by conditioned media (Patterson & Chun, 1977) or contact with killed cells (Hawrot, 1980). Cell-cell contact itself may provide positional information (McMahon, 1973). Connective tissue stroma can have inductive effects (Meier & Hay, 1974) as can hormones. Diffusion within enclosed extracellular spaces such as are found in many embryos, is another plausible route for relatively private communication. Furthermore, macromolecules can be passed between some kind of cells as from glia to axon (Lasek et al., 1974), and nerve growth factor appears to be an important trophic substance that would presumably be transmitted between cells extracellularly (Mobley et al., 1977).

Loss of coupling at a particular stage has been taken as a sign of completion of a determinative event; examples are given above. Yet cells can undergo mitosis as well as differentiate while coupled (Cavoto & Flaxman, 1974; O'Lague et al., 1970) or even while connected by cytoplasmic bridges (Woodruff & Telfer, 1980). If coupling mediates differentiation, then coupling between cells should be present while they are differentiating, i.e., the cells become different while coupled and because they are coupled. For this reason it does not contradict a role of gap junctions that there are fate maps with well defined borders where there appears to be no communication barrier (Jacobson, 1980). Furthermore, there is no necessity for coupling to disappear after the chemical interactions leading to pattern formation have occurred (Caveney, 1974). If the message is no longer sent or listened to, persistence of coupling is irrelevant or may serve a nutritional or different communicative function.

In summary, gap junctions are clearly not the only mechanism of intercellular communication in embryogenesis, although they can provide an intimate, regulatable and highly specific avenue of interaction. The possibility that they operate in some determinative processes seems increasingly likely as further mechanisms of modifiability and correlations with developmental changes are discovered. But a definite demonstration has yet to be made.

REFERENCES

Amsterdam, A. and J.D. Jamieson. 1974. Studies on dispersed pancreatic exocrine cells: I. Dissociation technique and morphologic characteristics of separated cells. J. Cell Biol. 63: 937-1056.

Albertini, D.F. and E. Anderson. 1974. The appearance and structure of intercellular connections during the ontogeny of the rabbit ovarian follicle with particular reference to gap junctions. J. Cell Biol. 53: 234-250.

Bennett, M.V.L. 1966. Physiology of electrotonic junctions. Ann. N.Y. Acad. Sci. 137: 509-539.

Bennett, M.V.L. 1973. Function of electrotonic junctions in embryonic and adult tissues. Fed. Proc. 32: 65-75.

Bennett, M.V.L. 1973. Permeability and structure of electrotonic junctions and intercellular movement of tracers. In:

"Intracellular Staining Techniques in Neurobiology". (Eds. S.B. Kater and C. Nicholson.) pp. 115-133, Elsevier, New York.

Bennett, M.V.L. 1978. Junctional permeability. In: "Intercellular Junctions and Synapses". (Eds. J. Feldman, N.B. Gilula and J.D. Pitts.) pp. 25-36, Chapman and Hall, London.

Bennett, M.V.L., J.E. Brown, A.L. Harris and D.C. Spray. 1978. Electrotonic junctions between Fundulus blastomeres: Reversible block by low intracellular pH. Biol. Bull. 155: 442.

Bennett, M.V.L., and D.A. Goodenough. 1978. Gap junctions, electrotonic coupling, and intercellular communication. Neurosci. Res. Prog. Bull. 16: 373:486.

Bennett, M.V.L., M.E. Spira and G.D. Pappas. 1972. Properties of electrotonic junctions between embryonic cells of Fundulus. Devel. Biol. 29: 419-435.

Bennett, M.V.L., M.E. Spira and D.C. Spray. 1978b. Permeability of gap junctions between embryonic cells of Fundulus: A reevaluation. Devel. Biol. 65: 114-128.

Bennett, M.V.L., D.C. Spray and A.L. Harris. 1981. Electrical coupling in development. Amer. Zool. 21: 413-427.

Browne, C.L., H.S. Wiley and J.N. Dumont. 1979. Oocyte-follicle cell gap junctions in Xenopus laevis and the effects of gonadotropin on their permeability. Science 203: 182-185.

Carew, T.J. and E.R. Kandel. 1976. Two functional effects of decreased conductance EPSPs: synaptic augmentation and increased electrotonic coupling. Science 192: 150-153.

Cavney, S. 1974. Intercellular communication in a positional field. Movement of small ions between insect epidermal cells. Devel. Biol. 40: 311-322.

Cavoto, F.V. and B.A. Flaxman. 1973. Low-resistance pathways between mitotic and interphase epidermal cells in vitro. J. Cell Biol. 58: 223-225.

Cooke, J. 1980. Specification in the developing eye for orientation of mapping to the brain. Trends Neurosci. 3: 45-48.

Decker, R.S. 1976. Hormonal regulation of gap junction differentiation. J. Cell Biol. 69: 669-685.

DeMello, W.C. 1975. Effect of intracellular injection of calcium and strontium on cell communication in heart. J. Physiol. 250: 231-245.

Duguid, J. and J.P. Revel. 1976. The protein components of the gap junction. Cold Spring Harbor Symp. Quant. Biol. 40: 45-47.

Elias, P.M. and Friend, D.S. 1976. Vitamin-A-induced mucous metaplasia. An in vitro system for modifying tight and gap junction differentiation. J. Cell Biol. 68: 173-188.

Epstein, M., J.D. Sheridan and R.G. Johnson. 1977. Formation of low-resistance junctions in vitro in the absence of protein synthesized ATP production. Exp. Cell Res. 104: 25-30.

Fallon, R.F. and D.A. Goodenough. 1981. Five-hour half-life of mouse liver gap-junction protein. J. Cell Biol. 90: 521-526.

Finbow, M., S.B. Yancey, R. Johnson, and J.P. Revel. 1980. Independent lines of evidence suggesting a major gap junctional protein with a molecular weight of 26,000. Proc. Natl. Acad. Sci., U.S.A. 77: 970-974.

Fishbach, G.D., Henkart, M.P., Cohen, S.A., J. Breuer, J. Whysner and F.M. Neal. 1974. Studies on the development of neuromuscular junctions in cell culture. In: "Synaptic Transmission and Neuronal Interaction". (Ed. M.V.L. Bennett.) pp. 259-283, Raven Press, New York.

Furshpan, E.J. and D.D. Potter. 1968. Low-resistance junctions between cells in embryos and tissue culture. Curr. Topics Devel. Biol. 3: 95-127.

Garfield, R.E., M.S. Kanman and E.E. Daniel. 1980. Gap junction formation in myometrium: control by estrogens, progesterone and prostaglandins. Am. J. Physiol. 238: C81-89.

Gierer, A. 1977. Biological features and physical concepts of pattern formation exemplified by hydra. Curr. Topics Devel. Biol. 11: 16-59.

Gilula, N.B. 1977 Gap junctions and cell communication. In: "International Cell Biology". (Eds. B. Brinkley and K. Porter.) pp. 61-69, Rockefeller Univ. Press, New York.

Gilula, N.B., O.R. Reeves and A.B. Steinbach. 1972. Metabolic coupling, ionic coupling and cell contacts. Nature 235: 262-265.

Ginzberg, R.D. and N.B. Gilula. 1979. Modulation of cell junctions during differentiation of the chicken otocyst sensory epithelium. Devel. Biol. 68: 110-129.

Goodenough, D.A. and N.B. Gilula. 1974. The splitting of hepatocyte junctions and zonulae occludentes with hypertonic disaccharides. J. Cell Biol. 61: 575-590.

Goodman, C.S. and Spitzer, N.C. 1979. Embryonic development of identified neurons: Differentiation from neuroblast to neurone. Nature 280: 208-214.

Griepp, E.B. and M. Bernfield. 1978. Acquisition of synchronous beating between embryonic heart cell aggregates and layers. Exp. Cell Res. 113: 263-272.

Grimmelikhuijzen, C.J.P. and H.C. Schaller. 1979. Hydra as a model organism for the study of morphogenesis. Trends Biochem. Sci. 4: 265-276.

Hanna, R.B., P.G. Model, D.C. Spray, A.L. Harris and M.V.L. Bennett. 1980. Gap junctions in early amphibian embryo. Amer. J. Anat. 158: 111-114.

Hanna, R.B., G.D. Pappas and M.V.L. Bennett. 1972. Structural changes associated with increased coupling resistance in the septate axon of the crayfish. Biol. Bull. 157: 370.

Hanna, R.B., T.S. Reese, R.L. Ornberg, D.C. Spray and M.V.L. Bennett. 1981. Fresh-frozen gap junctions: Structural detail in coupled and uncoupled states. J. Cell Biol. 91: 125a.

Hanna, R.B., D.C. Spray, P.G. Model, A.L. Harris and M.V.L. Bennett. 1978. Ultrastructure and physiology of gap junctions of an amphibian embryo; effects of CO_2. Biol Bull. 155: 442.

Harris, A.L., D.C. Spray and M.V.L. Bennett. 1981. Kinetic properties of a voltage dependent junctional conductance. J. Gen. Physiol. 77: 95-117.

Hertzberg, E.L. 1980. Biochemical and immunological approaches to the study of gap junctional communication. In Vitro 16: 1057-1067.

Hertzberg, E.L. and N.B. Gilula. 1979. Isolation and characterization of gap junctions from rat liver. J. Biol. Chem. 254: 2138-2147.

Hess, R. and R. Weingart. 1980. Intracellular free calcium modified by pH_i in sheep cardiac Purkinje fibres. J. Physiol. 307: 60S-61P.

Heuser, J.E., T.S. Reese and D.M.D. Landis. 1975. Preservation of synaptic structure by quick freezing. Cold Spring Harbor Symp. Quant. Biol. 40: 17-24.

Hunt, R.K. and M. Jacobson. 1974. Neuronal specificity revisited. Curr. Topics Devel. Biol. 8: 203-259.

Iwatsuki, N. and O.H. Petersen. 1979. Pancreatic acinar cells: The effect of carbon dioxide, ammonium chloride, and acetylcholine on intercellular communication. J. Physiol. 291: 317-326.

Jacobson, M. 1980. Clones and compartments in the vertebrate central nervous system. Trends Neurosci. 3: 3-5

Johnson, R., J. Hammer, J. Sheridan and J.P. Revel. 1974. Gap junction formation between reaggregated Novikoff hepatoma cells. Proc. Natl. Acad Sci., U.S.A. 71: 4536-4540.

Johnston, M.F., S.A. Simon and F. Ramon. 1980. Interaction of anesthetics with electrical synapses. Nature 286: 498-500.

Karlin, A. 1980. Molecular properties of nicotinic acetylcholine receptors. Cell Surface Rev. In press.

Kelly, A.M. and S.J. Zacks. 1969. The fine structure of motor endplate morphogenesis. J. Cell Biol. 42: 154-169.

Landmesser, L. and G. Pilar. 1972. The onset and development of transmission in the chick ciliary ganglion. J. Physiol. 222: 691-713.

Lane, N.J. and L.S. Swales. 1980. Dispersal of junctional particles, not internalization, during in vivo disappearance of gap junctions. Cell 19: 579-586.

Lasek, R.J., J.H. Gainer and R.J. Przybylski. 1974. Transfer of newly synthesized proteins from Schwann cells to the squid giant axon. Proc. Nat. Acad. Sci. 71: 1188-1192.

Lawrence, T.S., W.H. Beers and N.B. Gilula. 1978. Transmission of hormonal stimulation by cell-to-cell communication. Nature 272: 501-506.

Lo, C.W. and N.B. Gilula. 1980. Gap junctional communication in the post-implantation mouse embryo. Cell 18: 411-422.

LoPresti, V., E.R. Macagno and C. Levinthal. 1974. Structure and development of neuronal connections in isogenic organisms: Transient gap junctions between growing optic axons and lamina neuroblasts. Proc. Nat. Acad. Sci. 71: 1098-1102.

Mackie, G.O. 1980. Jellyfish neurobiology since Romanes. Trends Neurosci. 2: 13-16.

Makowski, L., D.L.D. Caspar, W.C. Phillips and D.A. Goodenough. 1977. Gap junction structure. II. Analysis of the X-ray diffraction data. J. Cell Biol. 74: 629-645.

MacMahon, D. 1973. A cell contact model for position determination in development. Proc. Nat. Acad. Sci., U.S.A. 70: 2396-2400.

Meda, P., A. Perrelet and L. Orci. 1979. Increase in gap junctions between pancreatic β-cells during stimulation of insulin secretion. J. Cell Biol. 82: 441-448.

Meier, S. and E.D. Hay. 1974. Control of differentiation by extracellular materials. Collagen as a promotor and stabilizer of epithelial stroma production. Devel. Biol. 38: 249-270.

Merk, F.B. and N.S. McNutt. 1972. Nexus junctions between dividing and interphase granulosa cells of the rat ovary. J. Cell Biol. 55: 511-515.

Messenger, E.A. and A.E. Warner. 1979. The function of the sodium pump during differentiation of amphibian embryonic neurons. J. Physiol. 292: 85-105.

Mobly, W.C., A.C. Server, D.N. Ishii, R.R. Riopelle and E.M. Shooter. 1977. Nerve growth factor. New Eng. J. Med. 297:1149-1158.

Ne'eman, Z., M.E. Spira and M.V.L. Bennett. 1980. Formation of gap and tight junctions between reaggregated blastomeres of the killifish, Fundulus. Amer. J. Anat. 158: 251-262.

Neher, E. and C.F. Stevens. 1977. Conductance fluctuations and ionic pores in membranes. Ann. Rev. Biophys. Bioeng. 6: 345-372.

Nicolson, B.J., M.W. Hunkapiller, L.E. Hood and J.P. Revel. 1980. Partial sequencing of the gap junctional protein from rat lens and liver. J. Cell Biol. 87: 200a.

O'Lague, P., H. Dalen, H. Rubin and C. Tobias. 1970. Electrical coupling: Low-resistance junctions between mitotic and interphase fibroblasts in tissue culture. Science 170: 464-466.

Page, E. and Y. Shibata. 1981. Permeable junctions between cardiac cells. Ann. Rev. Physiol. 43: 431-441.

Patterson, P.H. and L.L.Y. Chun. 1977. The induction of acetylcholine synthesis in primary cultures of dissociated rat sympathetic neurons. I. Effects of conditioned medium. Devel. Biol. 56: 263-280.

Perrachia, C. 1980. Structural correlates of gap junction permeation. Int. Rev. Cytol. 66: 81-146.

Pitts, J.C. and J.W. Simms. 1977. Permeability of junctions between animal cells. Intercellular transfer of nucleotides but not of macromolecules. Exp. Cell Res. 104: 153-163.

Raviola, E., D.G. Goodenough and G. Raviola. 1981. Structure of rapidly frozen gap junctions. J. Cell Biol. 87: 273-279.

Revel., J.P. 1978. Morphological and chemical organization of gap junctions. In: "Ninth International Congress on EM", Vol. III, (Ed. J.M. Sturgess,) pp. 651-672.

Revel, J.P. and D.A. Goodenough. 1970. Cell coats and intercellular matrix. In: "Chemistry and Molecular Biology of Intercellular Matrix". (Ed. E.A. Balasz.) pp. 1361-1369, Academic Press, New York.

Revel, J.P., E.B. Griepp, M. Finbow and R. Johnson. 1978. Possible steps in gap junction formation. Zool. 6: 139-144.

Revel, J.P. and M. Karnovsky. 1967. Hexagonal arrays of subunits in intercellular junctions of the mouse heart and liver. J. Cell Biol. 33: 7-12.

Rink, T.J., R.Y. Tsien and A.E. Warner. 1980. Free calcium in Xenopus embryos measured with ion-selective microelectrodes. Nature 283: 658-660.

Robertson, J.D., G. Zampighi and S.A. Simon. 1981. Biophysical studies of mammalian lens junctions. Biophys. J. 33: 77a.

Rose, B. and W.R. Loewenstein. 1975. Permeability of cell junction depends on local cytoplasmic calcium activity. Nature 254: 250-252.

Sheridan, J. 1978. Junction formation and experimental modification. In: "Intercellular Junctions and Synapses". (Eds., J. Feldman, N.B. Gilula and J.D. Pitts.) pp. 37-60, Chapman and Hall, London.

Simpson, I., B. Rose and W.R. Loewenstein. 1977. Size limit of molecules permeating the junctional membrane channels. Science 195: 294-296.

Sotolo, C., R. Llinas and R. Baker. 1974. Structural study of inferior olivary nucleus of the cat: Morphological correlates of electrotonic coupling between neurons. Brain Research 37: 294-300.

Spira, M.E. and M.V.L. Bennett. 1972. Synaptic control of electrotonic coupling between neurons. Brain Research 37:

Spira, M.E., D.C. Spray and M.V.L. Bennett. 1980. Synpatic organization of expansion motoneurons of Navanax inermis. Brain Research 195: 241-269.

Spitzer, N.C. 1980. Electrical uncoupling of vertebrate spinal cord neurons during development. Soc. Neurosci. Abstr. 6: 287.

Spray, D.C., A.L. Harris and M.V.L. Bennett. 1979. Voltage dependence of junctional conductance in early amphibian embryos. Science 204: 432-434.

Spray, D.C., A.L. Harris and M.V.L. Bennett. 1981a. Gap junctional conductance is a simple and sensitive function of intracellular pH. Science 211: 712-715.

Spray, D.C., A.L. Harris and M.V.L. Bennett. 1981b. Equilibrium properties of a voltage dependent junctional conductance. J. Gen. Physiol. 77: 77-93.

Spray, D.C., A.L. Harris, R.L. White and M.V.L. Bennett. 1981c. Glutaraldehyde differentially affects gap junctional conductance and its pH and voltage dependence. Biophys. J. 33: 108a.

Spray, D.C., A.L. Harris, R.L. White and M.V.L. Bennett. 1981d. Pharmacologically distinct sites mediate voltage and pH dependence of gap junctional conductance. Int. Biophys. Congr. Abst., p. 155.

Spray, D.C., J.H. Stern, A.L. Harris and M.V.L. Bennett. 1982a. Gap junctional conductance: Comparison of sensitivities to H and Ca ions. Proc. Nat. Acad. Sci. U.S.A., in press.

Spray, D.C., A.L. Harris and M.V.L. Bennett. 1982b. Comparison of pH and Ca dependence of gap junctional conductance. In: "Intracellular pH". (Eds. R. Nuccitelli and D. Deamer.) Alan R. Liss, New York, in press.

Staehelin, L.A. 1974. Structure and function of intercellular junctions. Int. Rev. Cytol. 39: 191-274.

Stern, J.H., D.C. Spray, A.L. Harris and M.V.L. Bennett. 1980. Gap junctions: Quantitative comparison of reduction in conductance by H and by Ca ions in a perfused preparation. Biol. Bull. 159: 493.

Stewart, W.W. 1981. Lucifer dyes - highly fluorescent dyes for biological staining. Nature 292: 17-21.

Turin, L. and A.E. Warner. 1978. Carbon dioxide reversibly abolishes ionic communication between cells of early amphibian embryo. Nature 270: 56-57.

Tsien, R.W. and R. Weingart. 1976. Inotropic effect of cyclic AMP in calf ventricular muscle studied by a cut end method. J. Physiol. 260: 117-141.

Unwin, P.N.T. and G. Zampighi. 1980. Structure of the junction between communicating cells. Nature 283: 545-549.

Warner, A.E. 1973. The electrical properties of the ectoderm in the amphibian embryo during induction and early development of the nervous system. J. Physiol. 235: 267-286.

Warner, A.E. 1979. Consequences of cell to cell signals in the early embryo. In: "The Role of Intercellular Signals: Navigation, Encounter, Outcome". (Ed. J.G. Nicholls,) pp. 179-200. Verlag Chemie, Berlin.

Weingart, R., P. Hess and W.R. Reber. 1981. Influence of intracellular pH on cell-to-cell coupling in sheep Purkinje fibers. In: "International Symposium on Normal and Abnormal Conduction in the Heart." Eds. H. Hoffman, M. Lieberman and A. Paes de Carvalho. Elsevier Publ. Co., Amsterdam, in press.

Williams, E.H. and R.L. Dettaan. 1981. Electrical coupling among heart cells in the absence of ultrastructurally defined gap junctions. J. Membr. Biol. 60: 237-248.

Wolpert, L. 1978. Gap junctions: Channels for communication in development. In: "Intercellular Junctions and Synapses". (Eds. J. Feldman, N.B. Gilula and J.C. Pitts,) pp. 83-94. Chapman and Hall, London.

Wolpert, L., A. Hornbruck and M.R.B. Clark. 1974. Positional information and positional signalling in hydra. Amer. Zool.

14: 647-663.

Wood, R.L. 1979. The fine structure of the hypostome and mouth of hydra. II. Transmission electron microscopy. Cell Tissue Res 199: 319-338.

Woodruff, R.I. and Telfer, W.H. 1980. Electrophoresis of proteins in intercellular bridges. Nature 286: 84-86.

Yee, A.G. and J.P. Revel. 1978. Loss and reappearance of gap junctions in regenerating liver. J. Cell Biol. 78: 554-564.

Zipser, B. 1979. Voltage-modulated membrane resistance in coupled leech neurons. J. Neurophysiol. 42: 465-475.

Transient Dye Coupling Between Developing Neurons Reveals Patterns of Intercellular Communication During Embryogenesis

Jonathan A. Raper and Corey S. Goodman

Because their differentiated forms are very complex and highly interrelated, neurons are likely to require considerable information about their environment as they develop. Each neuron acquires very specific biophysical, biochemical, and morphological properties, and a specific set of synaptic connections which are essential for its ultimate functioning. For example, each neuron sends axonal processes over large distances, upon very precise pathways, to reach specific targets (eg Lance-Jones and Landmesser, 1981; Raper et al, 1982a,b). The growth cones that accomplish these tasks must interact with their environment in such a way that they can distinguish the proper pathways upon which to elongate, the proper locations at which to branch, and the proper targets upon which to synapse. These considerations imply that some form of cell to cell communications must play a decisive role in neuronal development.

Intercellular communication can in principle be accomplished in any number of ways. For example, receptors on a cell's surface could recognize membrane bound, extracellular matrix bound, or diffusible molecules in the environment. Alternatively, diffusible substances could be exchanged between the interiors of cells whose cytosols are connected in some way. The ubiquity of electrical coupling and gap junctions between the interiors of a wide variety of cells in developing tissues (e.g. Potter et al, 1966; Blackshaw and Warner, 1976; Lopresti et al, 1974; Goodman and Spitzer, 1979) suggests that some intercellular communication could occur by this mechanism.

While studying the development of neurons in the central nervous system of the grasshopper, we have noticed distinctive patterns of coupling between the interiors of individually

identified neurons at early stages of their differentiation.   If
the fluorescent dye Lucifer Yellow (Stewart, 1978) is
iontophoresed into single embryonic neurons through a
microelectrode, dye is observed to spread in reproducible patterns
to other neurons.   The observation that (i) molecules more
substantial than inorganic ions can (ii) travel between particular
subsets of neurons (iii) during important developmental times
strengthens the supposition that important information could be
exchanged between the interiors of developing neurons.

The grasshopper is a particularly favorable preparation in
which to study how particular neurons acquire their complex
differentiated forms (Goodman and Spitzer, 1979; Goodman et al,
1979; Goodman and Bate, 1981).   A great deal of work in adults has
generated a catalogue of individually identified neurons whose
morphological, physiological, and functional properties are known
(e.g. Burrows, 1973; Pearson et al, 1980).   Coincidentially, the
progenitor cells that give rise to the central nervous system of
the grasshopper are large, accessible, and have been actively
studied for some time (Wheeler, 1893).   Each progenitor cell can
be individually identified from embryo to embryo (Bate, 1976).
Moreover, cell lineages can be constructed which relate a cell's
birth order from a particular progenitor cell to its mature
identified neuronal phenotype (Goodman and Spitzer, 1979; Goodman
et al, 1981; Goodman and Bate, 1981).   These characteristics allow
the differentiation of individually identifiable neurons to be
studied from their birth to their maturation.

The central nervous system of the grasshopper is composed of
reiterated segmental ganglia.   Each ganglion is composed of
neurons produced by a precise pattern of neuronal precursor cells
of two distinctive types called neuroblasts (NBs) (Bate, 1976) and
midline precursors (MPs) (Bate and Grunewald, 1981).   Most of the
neurons in each ganglion are produced by a plate of 61 neuroblasts
(Figure 1).   All but one of these neuroblasts sit on the ventral
surface of the ganglion.   Each neuroblast is a stem cell which
through a continuous process of asymmetric cell divisions gives
rise to a string progeny termed 'ganglion mother cells.'   Each
ganglion mother cell divides once to yield two post mitotic
ganglion cells which differentiate into neurons.   A string of
progeny cells is pushed dorsally from each neuroblast into the
center of the ganglion.   The neurons initiate axonal processes
which climb to the dorsal surface of the ganglion where they

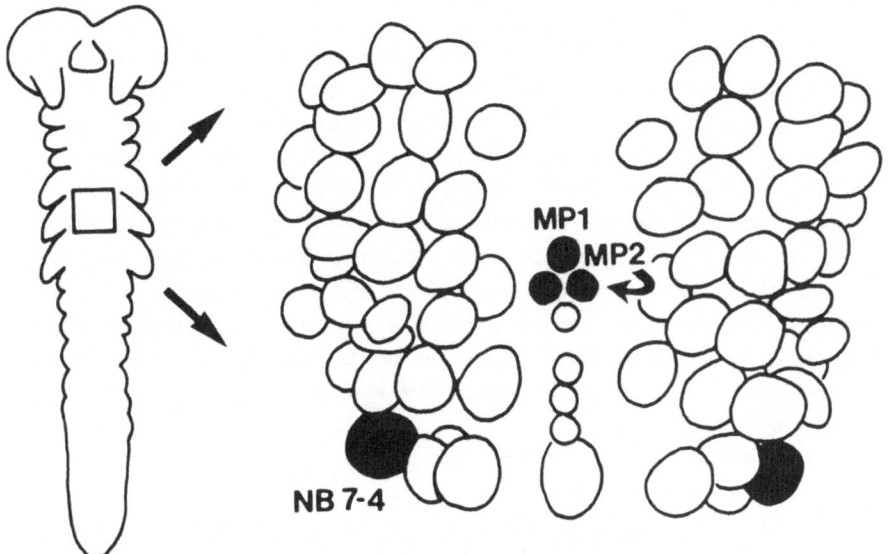

Figure 1. A camera lucida drawing of the neuroblasts (NBs) and
midline precursors cells (MPs) that generate the neurons in one
segmental ganglion of an embryonic grasshopper. The progeny of
MP1, MP2$_R$, and MP2$_L$ pioneer the interganglionic connectives. The
first neurons produced by the bilaterally symmetric NB 7-4s
pioneer the posterior connective.

choose their appropriate pathways and branching patterns in the
developing neuropil. Although most neurons are produced by
neuroblasts, some important neurons are produced by the second
class of precursor cells: the seven midline precursors (MP1,
2_L, 2_R, 3-6). These precursor cells lie near the midline of the
ganglion on its dorsal side. Each undergoes a single symmetric
cell division which gives rise to a pair of neuronal progeny.

Bate and Grunewald (1981) used serial electron and light
micrographs to trace the first axons in the interganglionic
connectives back to the progeny of the MP1 and MP2 progenitor
cells (shaded in Figure 1). Each MP2 divides once to give rise to
a ventral (vMP2) and a dorsal (dMP2) daughter cell. The vMP2
extends an ipsilateral axon anteriorly. The dMP2 extends an
ipsilateral axon posteriorly. The single MP1 progenitor gives
rise to a pair of bilaterally symmetric daughters, each of which
comes to lie dorsal to the MP2 progeny. The vMP2, dMP2, and MP1
thus form a trio of cell bodies on each side of the ganglion.

88

Each MP1 daughter extends an ipsilateral axon posteriorly in close association with the dorsal MP2 axon originating on the same side.

If living segmental ganglia are examined from their dorsal side with Nomarski optics, the very first growth cones and axons forming the ganglionic commissures and interganglionic connectives can be visualized. We have visually traced the growth cones which pioneer the posterior commissure back to their cell bodies at the posterior, lateral edges of each ganglion. Their identities were then confirmed by filling their cell bodies with Lucifer Yellow and visualizing their growth cones with the fluorescence microscope. In this way it is possible to show that the posterior commissure of each ganglion is pioneered by the first neurons to arise from the identified, bilaterally symmetric neuroblasts 7-4 (shaded in Figure 1). The axonal projections of the cells which pioneer the posterior commissure and the interganglionic connnectives are diagramed in Figure 2.

Initially, neuroblasts are dye coupled to each other, and to their young progeny. In time they uncouple from each other, yet remain strongly coupled to their own young progeny (Goodman and

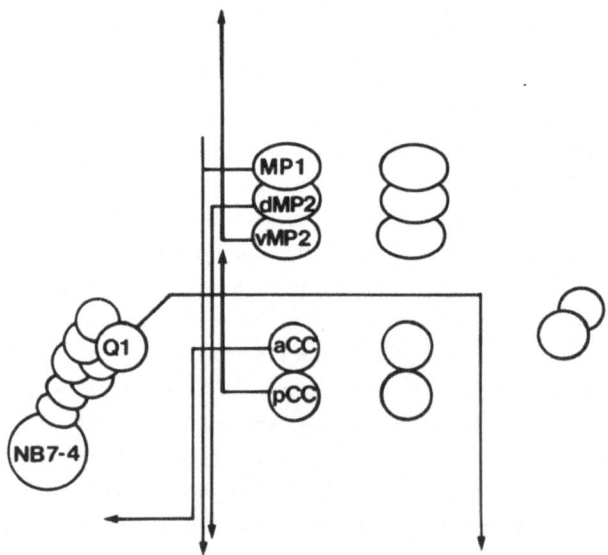

Figure 2. A schematic of the routes taken by the axons of some of the earliest differentiating neurons in a segmental ganglion. Anterior is to the top of the page in all figures.

Spitzer, 1979). However, at the time when a neuron first extends its growth cone, it becomes dye uncoupled from its neuroblast and from its younger axonless siblings. Thus, the initial pattern of dye coupling between clonally related cells disappears. We were interested to see that this primary coupling is replaced by a specific temporal and spatial pattern of secondary dye coupling that links reproducible groups of specific related and unrelated neurons.

If any cell is filled with Lucifer Yellow while its growth cone is actively extending, the injected dye does not spread back to its precursor but rather spreads into other neurons in a reproducible and distinctive pattern. An example is shown in Figure 3A. The cell Q1 which pioneers the posterior commissure was filled at the stage when its axonal process reached about half way to the ganglionic midline. The injected dye spread into four other neurons on the same side of the ganglion as the injected cell. Among the coupled cells are the ipsilateral MP1 and dMP2 cells. Also coupled are a pair of cells aCC and pCC, lying near the ipsilateral junction of the posterior commissure and the connective. When the MP1 is filled with Lucifer at the same developmental stage, dye spreads into the ipsilateral dMP2, vMP2, aCC, pCC, and Q1 cells (Figure 3B). It should be stressed that there are many neuronal cell bodies surrounding the Q1, aCC, and pCC cells which are not dye coupled to either the MP1 or Q1 cells.

More extensive patterns of dye coupling are revealed by filling Q1 or MP1 slightly later in development. Once Q1's axon grows a bit further toward the midline, its sibling Q2 initiates an axonal process and Q2's growth cone elongates along Q1's axon. Although their primary dye coupling via their cell bodies had disappeared before Q1 first initiated its axon, the two sibling neurons now develop secondary dye coupling via their growth cones and axons. Once Q1's axon reaches the ganglionic midline it meets and continues growing along the mirror symmetric process of its contralateral homologue. Then dye coupling from Q1 is no longer confined to the side of the ganglion ipsilateral to the injected cell (Figure 4). Dye spreads into the contralateral homologues of the unrelated cells that are coupled ipsilaterally. Q1 is observed to be coupled to (i) its immediate, ipsilateral sibling Q2 and the next two neuronal progeny of NB 7-4, (ii) its contralateral homologue Q1 (iii) the contralateral homologue of its immediate sibling Q2, and (iv) unrelated, bilateral neurons

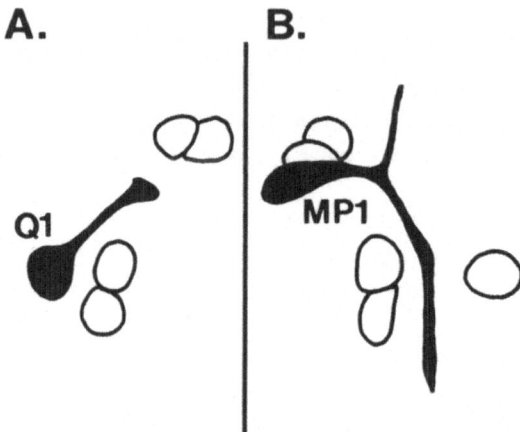

Figure 3. Dye coupling between embryonic neurons. (A) Lucifer Yellow was injected into the neuron, Q1 (shaded cell and process), which pioneers the posterior commissure. The injected dye spreads into four ipsilateral neurons (unshaded cell bodies). Dye did not cross the midline. (B) In the same ganglion Lucifer was injected into one of the neurons, MP1 (shaded cell and process), which pioneers the interganglionic connectives. The dye spread into the five cell bodies shown. For clarity, in this and following figures the axonal processes of coupled cell bodies are not drawn.

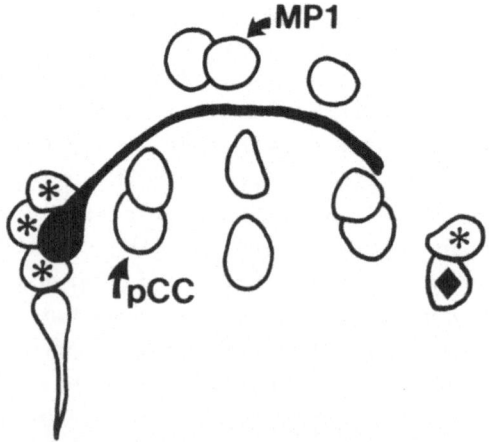

Figure 4. Dye coupling from Q1 after its process has crossed the ganglionic midline. Q1 (darkly shaded) is coupled to its (i) contralateral homologue Q1 (closed diamond), (ii) ipsilateral and contralateral siblings which arose from NB 7-4 just subsequent to its own birth (asterisks), and (iii) particular unrelated bilateral neurons (unmarked cell bodies).

including the MP1, aCC and pCC cells.

Additional patterns are revealed by filling MP1 after its axon leaves its ganglion of origin. Figure 5 shows a Lucifer Yellow fill of an MP1 whose axon has grown into the next posterior segment. Its axon joins the posterior running axon of its ipsilateral homologue. Coupling occurs from this MP1 to (i) its posterior ipsilateral homologue, and (ii) unrelated, segmentally repeated neurons including the pCC, Q1, and dMP2 cells. It should be stressed once again at this developmental time there are a large number of neuronal cell bodies, even in the vicinity of coupled cells, which are not coupled to the MP1. Additionally, many neurons have axons in the commissures and connectives near the Q1 and MP1 axons that do not become dye coupled to either cell. For example, we have not observed dye coupling between the dMP2 cell and the ipsilateral vMP2 residing in the next posterior segment, even though their axonal processes pass within a few microns of each other in the interganglionic connective and the filopodia of their growth cones overlap extensively.

DISCUSSION

Neurons Can Be Dye Coupled Through Their Axons.

The dye coupling we describe between embryonic neurons must generally involve their axonal processes since (i) the cell bodies of coupled cells are sometimes many ganglia apart, (ii) homologous neurons first become coupled around the time when their axons touch or begin to overlap, and (iii) previously uncoupled cells become coupled to their previously uncoupled siblings one by one, as each sibling initiates an axonal process.

Neurons Can Be Dye Coupled By Filopodial Processes.

Growth cones have numerous long, thin filopodia approximately 0.1 micron in diameter that extend for over 50 microns in many directions and contact large numbers of cells. One striking example that coupling can be mediated by the filopodia of an injected cell's growth cone is provided by the early interactions between the Q1 and aCC cells. Q1 is coupled to the aCC just as Q1's growth cone passes between the MP1 and aCC cells (Figure 3A). This can occur even if both (i) Q1's axonal process and growth cone are not in direct contact with the aCC cell and (ii) the aCC

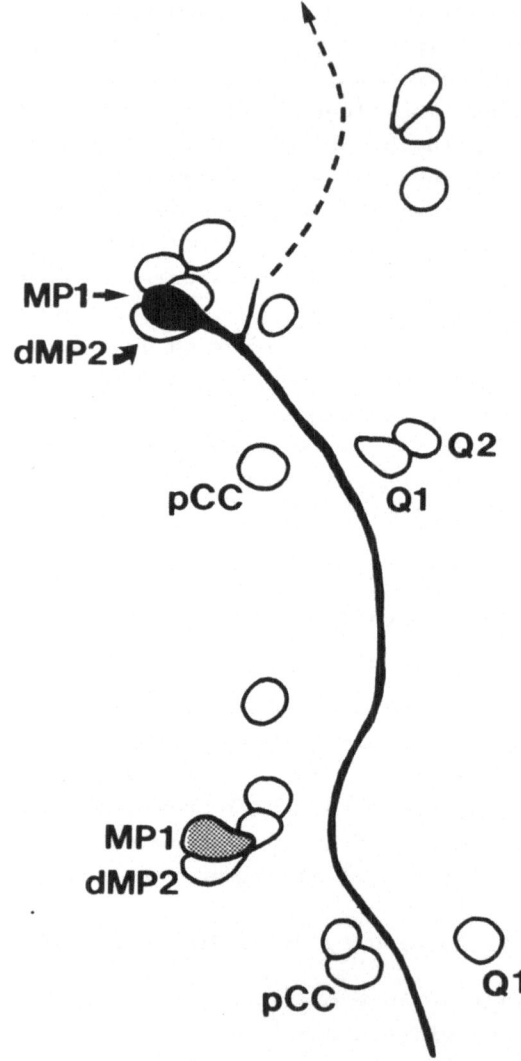

Figure 5. Dye coupling from MP1 after its process has grown through the next posterior ganglion. MP1 (darkly shaded) is coupled to specific cells including its (i) ipsilateral homologue MP1 (lightly shaded) and (ii) segmentally repeated unrelated cells including the MP2, pCC, and Q1 cells indicated.

cell has no axonal process of its own. The only dye containing
structures spanning the gap between these cells are the filopodia
of Q1's growth cone (see Taghert et al, 1982). At later
developmental times, as these and other cells remain coupled, the
profuse lateral filopodia that extend all along the lengths of
their axons could very well continue to mediate the dye cell
coupling we see at later times, or the axons themselves could
mediate the coupling.

The Axons Of Dye Coupled Neurons Tend To Be Travelling Companions.

     A general observation of some interest is that dye coupled
neurons tend to have axons that are in close contiguity for some
portion of their length. For example, the axons of the Q1 cell,
its contralateral homologue, and their associated sibling neurons
all travel together in the posterior commissure and are all dye
coupled. An example of the same phenomenon but between unrelated
neurons is provided by the coupled MP1 and dMP2 cells, whose axons
intertwine as they descend in the connectives. However, it is not
necessary for the region of close proximity to be particularly
long, since the coupled Q1 and pCC cells overlap for only a short
distance before they diverge and cross over one another. We
cannot routinely detect dye coupling between all neurons whose
axons fasiculate. For example, Raper, Bastiani, and Goodman
(1982b) show that the G and C neurons (which are the next two
neurons born from NB 7-4 after Q1 and Q2) fasiculate upon the
axons of four identified neurons in the neuropil. However, dye
coupling from either the G or the C cells to any of the four axons
upon which they fasiculate was only very rarely detected, and
coupling between some combinations was never detected.

Reciprocity Of Coupling.

     We find that dye coupling is generally reciprocal, in the
sense that any pair of coupled cells is coupled regardless of
which cell is filled with dye. Perhaps more significantly, at
earlier times in development filling either cell generally reveals
the same overall pattern of coupling. For example, in Figure 3A a
Q1 cell was filled and four other cells were found to be coupled
ipsilateral to the filled Q1 cell. No dye crossed the ganglionic
midline. Among the coupled cells was the MP1. The contralateral
homologue of the coupled MP1 was then filled with dye in the same

ganglion (Figure 3B). Coupling was once again confined to the
side of the ganglion ipsilateral to the newly injected cell. A
comparison of 3A and 3B demonstrate that a very similar pattern of
coupling was obtained in each case.

As neurons mature, and their growth cones diverge and
elongate, the overall patterns of coupling revealed by filling any
two cells which are themselves coupled tend to become increasingly
different. Thus, one could imagine that early in development
there are discrete non-overlapping subsets of neurons which are in
intimate communication with each other, but as development
proceeds, this form of communication interconnects members of many
separate subsets in new overlapping patterns of coupling.

No Alliance Lasts Forever.

Although dye coupling between neurons is ubiquitous at early
stages of embryogenesis, it is apparently absent in later embryos.
Lucifer does not spread between late stage siblings of
contralateral homologues of identified neurons which we know to be
coupled at earlier times and to our knowledge it has never been
observed between any adult grasshopper neurons in the segmental
ganglia. The coupling we describe in this paper appears to be a
transient phenomenon prevalent only during the times when neurons
are first differentiating and while their growth cones are
actively extending.

Specificity.

Coupling does not occur randomly between neurons. Specific
and stereotypic groups of neurons are coupled to a neuron at a
given time in development. Coupled cells need not have a similar
lineage. The only requirement is that their axons lie in close
proximity for at least a short distance. However, it is certainly
not the case that all axons that lie in close proximity are
coupled to one another. The axons of the dMP2 and the next
posterior vMP2 do not fasiculate but they lie within a few microns
of each other in the interganglionic connective, yet we have never
observed them to be dye coupled. A specific process of
cell-to-cell recognition is suggested by the finding that growth
cones and their filopodia contact many cells but only become
coupled to a specific subset of them.

Lucifer Yellow is a negatively charged naphthalimide dye that has a molecular weight of about 450 daltons. Its ability to pass between particular subsets of neurons at important times during their development reveals stereotypic connections between the interiors of these cells. We do not know what sort of membrane specializations are involved in the construction of these specific connections. Nor can we be sure that they are utilized for the exchange of developmentally meaningful information. Perhaps these issues can be addressed in the future by determining the structures responsible for dye coupling with electron micrographic studies, blocking coupling with an appropriate treatment to determine the developmental outcome, or removing one member of a coupled pair to see if its absence affects the subsequent differentiation of remaining cells.

ACKNOWLEDGEMENTS

We thank Michael Bate for his participation in some of our initial experiments. This work was supported by a NIH fellowship to J.A.R. and a NSF grant to C.S.G. who was also supported as an Alfred P. Sloan Research Fellow.

REFERENCES

Bate, C.M. 1976. Embryogenesis of an insect nervous system I. A map of the thoracic and abdominal neuroblasts in Locusta migratoria. J. Embryol. exp. Morph. 35: 107-123.

Bate, C.M. and E.B. Grunewald. 1981. Embryogenesis of an insect nervous system II. A second class of neuron precursor cells and the origin of the intersegmental connectives. J. Embryol. exp. Morph. 61: 317-330.

Blackshaw, S.E. and A.E. Warner. 1976. Low resistance junctions between mesoderm cells during development of trunk muscles. J. Physiol. 255: 209-230.

Burrows, M. 1973. Physiological and morphological properties of the metathoracic common inhibitory neuron of the locust. J. Comp. Physiol. 82: 59-78.

Goodman, C.S. and C.M. Bate. 1981. Neuronal development in the grasshopper. Trends Neurosci. 4, 163-169.

Goodman, C.S. and N.C. Spitzer. 1979. Embryonic development of identified neurons: Differentiation from neuroblast to neuron. Nature 280: 208-214.

Goodman, C.S., M. O'Shea, R.E. McCaman, and N.C. Spitzer. 1979. Embryonic development of identified neurons: Temporal pattern of morphological and biochemical differentiation. Science 204: 1219-1222.

Goodman, C.S., C.M. Bate and N.C. Spitzer. 1981. Embryonic development of identified neurons: origin and transformation of the H cell. J. Neurosci. 1: 94-102.

Lance-Jones, C. and L. Landmesser. 1981. Pathway selection by chick lumbosacral motoneurons during normal development. Proc. R. Soc Lond. in press

Lopresti, V., E.R. Macagno and C. Levinthal. 1974. Structure and development of neuronal connections in isogenic organisms: Transient gap junctions between growing optic axons and lamina neuroblasts. Proc. Nat. Acad. Sci. U. S. 71: 1098-1102.

Pearson, K.G., W.J. Keitler and J.D. Steeves. 1980. Triggering of locust jump by multimodal inhibitory interneurons. J. Neurophysiol. 43: 257-278.

Potter, D.D., E.J. Furshpan and E.S. Lennox. 1966. Connections between cells of the developing squid as revealed by electrophysiological methods. Proc. Nat. Acad. Sci. U. S. 55: 328-336.

Raper, J.A., M. Bastiani and C.S. Goodman. 1982a. Pathfinding by neuronal growth cones in grasshopper embryos: I. Divergent choices made by the growth cones of sibling neurons. J. Neurosci. in review.

Raper, J.A., M. Bastiani and C.S. Goodman. 1982b. Pathfinding by neuronal growth cones in grasshopper embryos: II. Selective fasciculation onto specific axonal pathways. J. Neurosci. in review.

Stewart, W.W. 1978. Functional connections between cells as revealed by dye-coupling with a fluorescent naphthalimide tracer. Cell 14: 741-759.

Taghert, P., M. Bastiani, R.K. Ho and C.S. Goodman. 1982. Guidance of pioneer growth cones: Filipodial contacts and coupling revealed with an antibody to Lucifer yellow. Science, in review.

Wheeler, W.M. 1893. A contribution to insect embryology. J. Morph. 8:1-160.

Molecular Biology of Lens Induction

W.J. van Doorenmaalen, H. van der Starre, P.T. Janssen, and M. van der Starre-van Bekkum

Lens induction can be considered one of those intercellular interactions in which tissue differentiation is started and growth is directed. Significant information on interactions during eye development has been contributed by the work of Coulombre (1965), Coulombre and Coulombre (1963), McKeehan (1951, 1958), Beebe et al.(1979), Genis-Galvez (1965), Genis-Galvez et al. (1967), Hay and Dodson, (1973), Niu and Twitty(1953), and Toivonen et al.(1976).

Chemical signals are assumed to be responsible for these tissue communications. The information might be transferred in the form of chemical substances that are secreted by the cells. Such secretions may be newly produced substances or metabolites that are continually produced. They may be specific for this inductive function or, as some investigators believe, they may not be specific at all. It is possible that changes in concentration of physiologically active chemicals or even deprivation of metabolic substances might be a signal or a message to cells in order to change their function. The matrix between cells is suspected to play an important role in tissue interactions since it is able to influence the passage or even the chemical structure of substances that pass from one cell or tissue to the other.

In our institute the lens has been a subject of investigation for many years. Primarily we have investigated the occurrence and distribution of the lens crystallins at embryological stages. For some time induction also has been a subject of research. The tissue interaction between the eye cup and the overlying ectoderm is a very suitable experimental model of which the accessibility, the very small amount of intercellular matrix, and the experience of the research group are the most important reasons for its choice.

.   In the literature there are ample indications that a mediator
emanating from the eye cup signals the overlying ectoderm to form
a lens placode. Niu and Twitty (1953) achieved differentiation of
gastrula-ectoderm by culturing this tissue in a medium that was
"conditioned" by dorsal lip tissue. Apparently dorsal lip has
excreted a morphogenetic substance into the medium that stimulates
gastrula ectoderm to differentiate. Sirlin and Brahma (1959)
demonstrated transfer of a compound from the optic cup to the
ectoderm by observing a transfer of [14C]-phenylalanine from the
eyecup into the developing lens. McKeehan (1958) placed
cellophane membranes between chick ectoderm and optic vesicle, and
prevented the differentiation of ectoderm except at the edges of
the membrane where reaction occurs at sites were ectoderm touched
eyecup. In contrast, differentiation occurred when agar strips
were used instead of cellophane (van der Starre, 1977, 1978).
Inserting Millipore filters of different pore sizes between optic
cup and ectoderm (Karkinen-Jääskeläinen, 1978a; van der
Starre, 1977) demonstrated the transfer of a molecule of certain
mass through the filter which was responsible for the induction.

     Several other investigators, such as Jacobson (1966),
Karkinen-Jääskeläinen (1978a,b), and Mizuno (1970, 1972)
doubt the specific and unique character of the inductor(s).
Jacobson (1966) found that amphibian lens induction is a result of
several tissue interactions during early embryogenesis. Initially
tissues such as heart mesoderm and endoderm take part while the
inductive role of the eye cup is considered to be the last, more
locally functioning, nonspecific stimulus.
Karkinen-Jääskiläinen (1978a,b) found lens induction in
mouse ectoderm initiated by the most bizzare tissues, such as
metanephros, salivary gland, lung tissue, mouse head mesenchyme,
etc. Mizuno (1970, 1972) was able to induce lenses in the chicken
as early as the epiblast stage by treatment with alcohol, or
denatured and fresh skin dermis (6 1/2 and 13 1/2 days) in the
obligatory presence of hypoblast. However, in our experiments we
were unable to detect any lens development in other than
eyecup/ectoderm combinations in the thousands of culture
experiments we have performed.

     In our institute, lens induction is investigated in vitro;
for full details on methods, refer to the papers of van der Starre
(1972, 1977, 1978). Optic cups of 72 hour chick embryos (stage

18) are cultured in organ culture dishes in a plasma clot containing Medium 199, embryo extract, and agar. Slices of agar 0.5 mm thick are placed over the eyecup and a sheet of head ectoderm from a stage 18 embryo is placed on the agar slice on the side opposite the optic cup. After several days of culture, the ectoderm is examined for the presence of a lens.

The van der Starres have investigated the tissues in the induction system under the most physiological conditions. Thus, they established a model that could be used as a control and to which the experimental data could be compared. A lens that developed in vitro is shown in Figure 1. The time of highest inductive activity by the eye cup could be established as well as the beginning and end of the inductive power; furthermore the responsiveness of the ectoderm could be determined (fig. 2). By the histological immunofluorescence technique, the presence of lens crystallins in the lens primordia could be demonstrated. As the culture technique has some negative aspects, the percentage of successful lens induction has been estimated by culturing many combinations of optic cup and ectoderm in all kinds of variations. It appeared that as skill increased during the years, the percentage of successful induction also increased to a steady number of 43%. The tissues appear to be harmed by the frequent manipulations, duration of preparation for culture, and by changes in temperature and ion concentration in the preparation medium. Greater skill decreases these circumstances and increases the number of inductions. Furthermore, all of the experiments were done in sufficient numbers to estimate the results statistically.

Van der Starre introduced a thin slice of agar between the optic cup and the ectoderm for two reasons. First, he wanted to

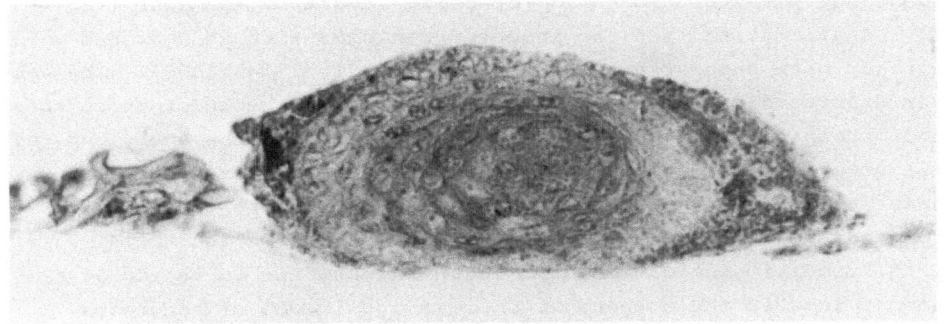

Fig. 1   Lens developed on an agar slice (AS)

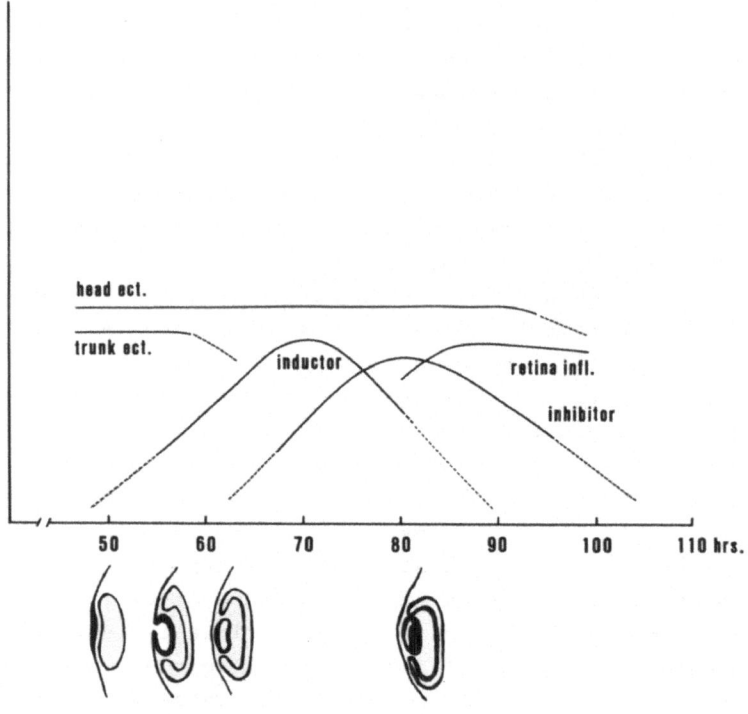

Fig. 2   Quantitative diagram of eyecup activity (inductor and inhibitor), retinal trophic influence and responsiveness of ectoderm (trunk and head) in relation to age.

exclude the possibility of lens development from remnants of ectoderm that were attached to the optic cup. In that case a lens would be found on that side of the agar where the lens is situated. The other reason was to investigate if there is a substance passing through and/or being caught by the agar (fig. 3). Confirmation that the inductor can pass through agar had already been demonstrated by McKeehan (1951, 1954, 1958). The van der Starres did a series of experiments using the information they had obtained on the time of greatest activity of the optic cup and the greatest susceptibility of the ectoderm. After a slice of agar was used between an eye cup and ectoderm, the tissues were removed and a new sheet of ectoderm placed on the agar (fig. 3.1). This "conditioned" or "induced" agar induced the formation of a second lens in the absence of the eye cup in 39% of the cases. The same slice of "conditioned" agar could be used with a new

sheet of ectoderm to induce a lens in the absence of optic cup
(figs. 3.1 and 2a).

This proves indisputably that a substance has been caught in
the agar, probably originating from the optic cup, which is
responsible for the information to the ectoderm and after which a
lens placode starts developing in the ectoderm. Also,
"conditioning" the agar by laying it on the optic cup without
ectoderm for at least 6 hours, gave the agar the ability to
initiate placode formation. Apparently the presence of ectoderm
is not obligatory for the eye cup to start "inductor" production
(fig. 3.2).

The species specificity of the inductor also has been
investigated. Using the same technique of placing a thin layer of
agar between the tissues, duck and chick were tested in
combinations with chicken, as seen in Figure 4. The ectoderm of
duck and quail also has a susceptibility gradient but it is
different from chick and specific for each of the species (van der
Starre, in press). Thus, cross reactivity has been proven among
birds. However. when chicken eye cups were cultured with ectoderm
of the mouse and vice versa no induction at all was observed.
Numerous cultures (± 150) were done with this combination, at

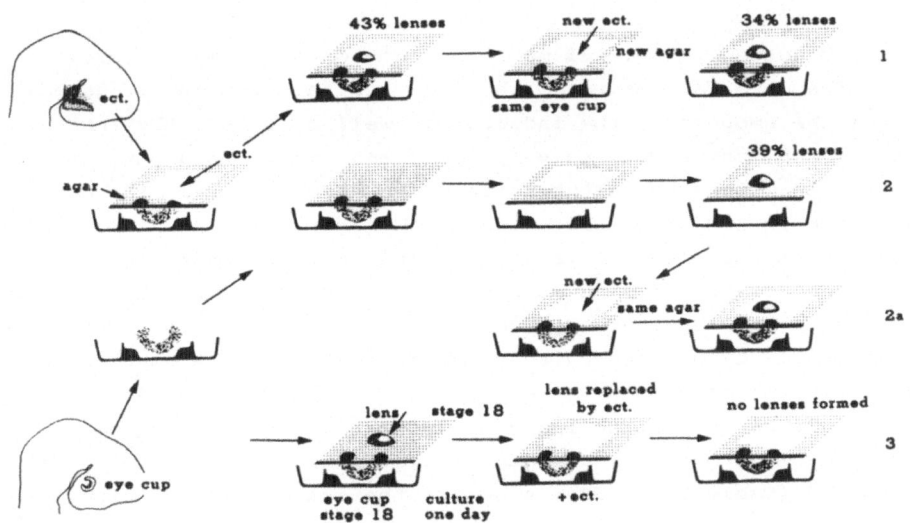

Fig. 3 Diagram of Culture Experiments. See text for explanation.

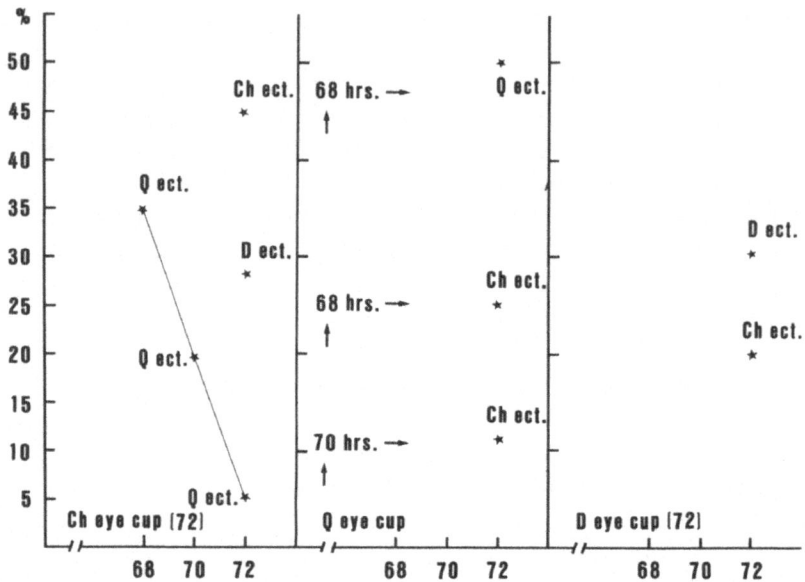

Fig. 4 Experimental results of heterologeous induction between
Quail (Q), Chick (Ch) and Duck (D). Percentage of lenses formed
is given on the vertical, ectoderm age in hours on the horizontal.
Left column shows chick eyecup (72 hrs) with Ch., Q. and D. ect,
middle gives quail eyecup (68 and 70 hrs) with Q. and Ch. ect. and
to the right duck eyecup (72 hrs) with D. and Ch. ect is seen.

various ages with several media but not one lens was formed.

     The chemical nature of the inductor has been investigated.
Since the amounts of the inductor as well as other substances
released by the tissue is very small and since the amount of
medium in which the cultures are grown also is small, the approach
has been to add test substances to the agar. It had already been
found that the inductor is quite stable and is not easy to remove
from the agar by rinsing or washing in fluids. The first step was
to see if enzyme treatments would destroy inductive activity
(Table I). Loss of activity after treatment of the agar slices
with trypsin or pepsin (by soaking in the enzyme solution) suggest
a protein nature of the inductor. The action of RNase and
Actinomycin suggest that synthetic activity of intact cells is
involved (Table II). This supports the results of our experiments
in which homogenized optic cups were used as an inductor with
negative results, whereas minced optic cups did initiate lens
placode formation, probably due to activity of whole cells.

TABLE I

Influence of Various Enzymes Introduced into the Agar Slice

Agar slices treated with:	Incubation time	Number of cultures	% lens
1 Hanks	1 hr	273	33
2 RNAse	1 hr	161	31
3 Trypsin	1 hr	76	20
4 Pepsin	1 hr	48	10

TABLE II

Influence of Inhibitors Introduced into the Agar Slice

Treated with Agar slices in:	Incubation time	Number of cultures	% lens
1 Actinomycin D	1.5 hr	64	0
2 Puromycin	1.5 hr	64	7
3 Eye cups (72 hrs) in actinomycin	1.5 hr	32	0
4 Eye cups (72 hrs) in puromycin	1.5 hr	40	
5 Ectoderm (72 hrs) in actinomycin	1.5 hr	48	0
6 Ectoderm (72 hrs) in puromycin	1.5 hr	48	7

As a result of our former work we possess very specific antisera against the lens crystallins and (FISC) of the chick (van Doorenmaalen et al., 1968; Brahma et al., 1971a,b). In order to investigate whether or not crystallins or proteinlike substances carrying the same serological determinants were involved in induction, the agar slices were soaked in several antisera and used in induction experiments. As is seen from Table III no inhibition was produced by the antisera; as a matter of fact the percentage of induction even increased, probably as a result of the higher protein content caused by the presence of antiserum. Thus, this kind of protein does not play any part in the induction process.

TABLE III

Influence of various antisera introduced in the agar slice

	Agar slices treated with	Incubation time	Number of cultures	% lens
1	Normal rabbit serum	1 hr	48	37
2	Anti Alpha crystallin serum	1 hr	188	46
3	Anti Beta crystallin serum	1 hr	32	44
4	Anti total lens serum	1 hr	70	33
5	Anti used medium eye cup	1 hr	72	0
6	Anti used medium heart	1 hr	42	36
7	Anti used medium agar slices	1 hr	96	0

Thanks to our increasing skill, quite a number of "conditioned" or "induced" agar slices could be harvested and assuming that the inductor should diffuse into the medium as it does into the agar, large amounts of medium in which optic cups had been cultured was collected. With a special method (Brahma et al., 1971a,b) rabbit antisera were produced against "conditioned" agar slices as well as against pooled medium. These antisera were added to agar slices between ectoderm and eyecup as was done with the anticrystallin serum. From Table III it can be seen that induction did not occur at all when either anti"induced" agar or anti"conditioned medium" serum were introduced into the agar slices; thus, the antibodies were acting against the inductor(s). This is the first experiment in which an antiserum completely inhibits induction and the action is limited to this tissue function without any interference in other tissue processes. This result strongly suggests that antibodies have been formed against the inductor.

To exclude the possibility that the antiserum against conditioned medium or induced agar constituents contained antibodies against other products that are important for differentiation, a control serum was produced against medium conditioned by heart tissue, which does not induce lenses. These antisera, introduced into the agar slices between eye cup and ectoderm, produced no effect on lens induction as seen in the percentages of successful cultures (Table III).

Having proof of the presence of antibodies in the antisera, all kinds of immunological tests were done to localize a precipitate of the antibody/inductor complex. By immunoelectrophoresis it is seen that several lines were produced. However, they were caused by the antibodies to the eggwhite proteins ovalbumin and conalbumin. These proteins seem to be secreted by the eye cup and ectoderm as well as by heart tissue during culture. An indication of the presence of the inductor was not found.

Immunochemical methods like the Laurel method (Osserman, 1960) did not give any information on the inductive substance. After application of the immunofluorescence method to histological sections, all constituents of the tissue were stained, in spite of absorption with eggwhite proteins, etc. These dissappointing results might be explained by results of the following chemicoanalytical investigations, performed with the help of Janssen (in press). Development of a culture system that produced the inductive substance allowed us to harvest large amounts of conditioned culture medium from cultures of eye cups. In spite of the large volumes, it is anticipated that only small amounts of inductor would be collected. An even more troublesome aspect is that the conditioned medium is expected to contain quite a number of cell metabolites of which the character and concentration is not known. In spite of the expected difficulties, extracts of the culture medium were fractionated by means of dialysis as well as ultrafiltration, using PM 10 and UM 2 membranes. All the fractions had to be tested by tissue culture for inductive activity. It appeared that the activity was limited to the fraction of 3000 - 10,000 Daltons. The low molecular mass of the inductor explains the disappointing results of the immunochemical methods as antigens of that mass usually do not precipitate.

Gel filtration on Sephadex G-50 superfine also was used to identify the inductor by filtration of the 3000 - 10,000 Dalton fraction of conditioned medium from at least 1000 eyecups. The elution pattern showed extinction throughout the whole separation range. The activity of each fraction was tested by numerous tissue cultures. The pooled fractions, as controls, were still able to induce lenses and just recently one fraction was isolated in which inductive activity could be found corresponding to a molecular mass of approximately 5000.

Thin layer chromatography also was used to investigate the

characteristics of the inductor. Fraction 1000 - 10,000 Dalton
from the conditioned medium was dansylated using the method of
Cross and Labouesse (1969). The reaction product was
chromatographed and compared with control chromatograms of tissues
such as heart, liver, lens, ectoderm, heart culture medium, etc.
A constant pattern could be recognized (fig. 5a). Figure 5b is a
chromatogram of the conditioned eyecup medium and spots which are
more or less specific for this conditioned medium are marked.
They are found in the upper corner of the TLC and are comprised
of three elements of which two are confluent in this picture
(arrows). Janssen assumed that an inductor which signals the
ectoderm should in some way combine with the ectoderm. Therefore,
he absorbed the "conditioned" or used medium with ectoderm tissue
of the proper age and investigated this absorbed "conditioned"
medium by thin layer chromatography. The conditioned eyecup
conditioned medium after absorption (fig. 5c) can be compared with
the unabsorbed medium (fig. 5b). The three characteristic spots
are missing in the absorbed medium and might give some additional
structural information about the inductor. We are continuing our
efforts to purify and identify this substance but it is clear that
it will be at the cost of much labor and time.

A last interesting phenomenon of this developmental system has
to be mentioned. It was clear that eyecups cultured in vitro had
a longer period of inductor productivity than eye cups have in
vivo. We thought that probably an inhibition by other tissues
might be responsible for that phenomenon. The first place to look
for an inhibitor is in the lens. Therefore the following
experiment was performed. Eyecups were cultured beneath a sheet
of agar with a lens positioned on top. After an appropriate time
in culture, the "conditioned" agar only was cultured with the
ectoderm on top. No lens formation by the ectoderm could be
observed (figs. 2 & 3). The same eyecups, after having been
inhibited by the lens, begin to produce the inductor again some
time after the lens has been removed. This adds certainty to the

Fig. 5   Thin layer chromatogram of dansylated 1000 - 10,000
Dalton fraction of:
a)   heart tissue extract
b)   conditioned eyecup medium, arrows at "specific" spots
c)   conditioned eyecup medium, two times absorbed with ectoderm
tissue (72 hrs). "specific" spots are gone.

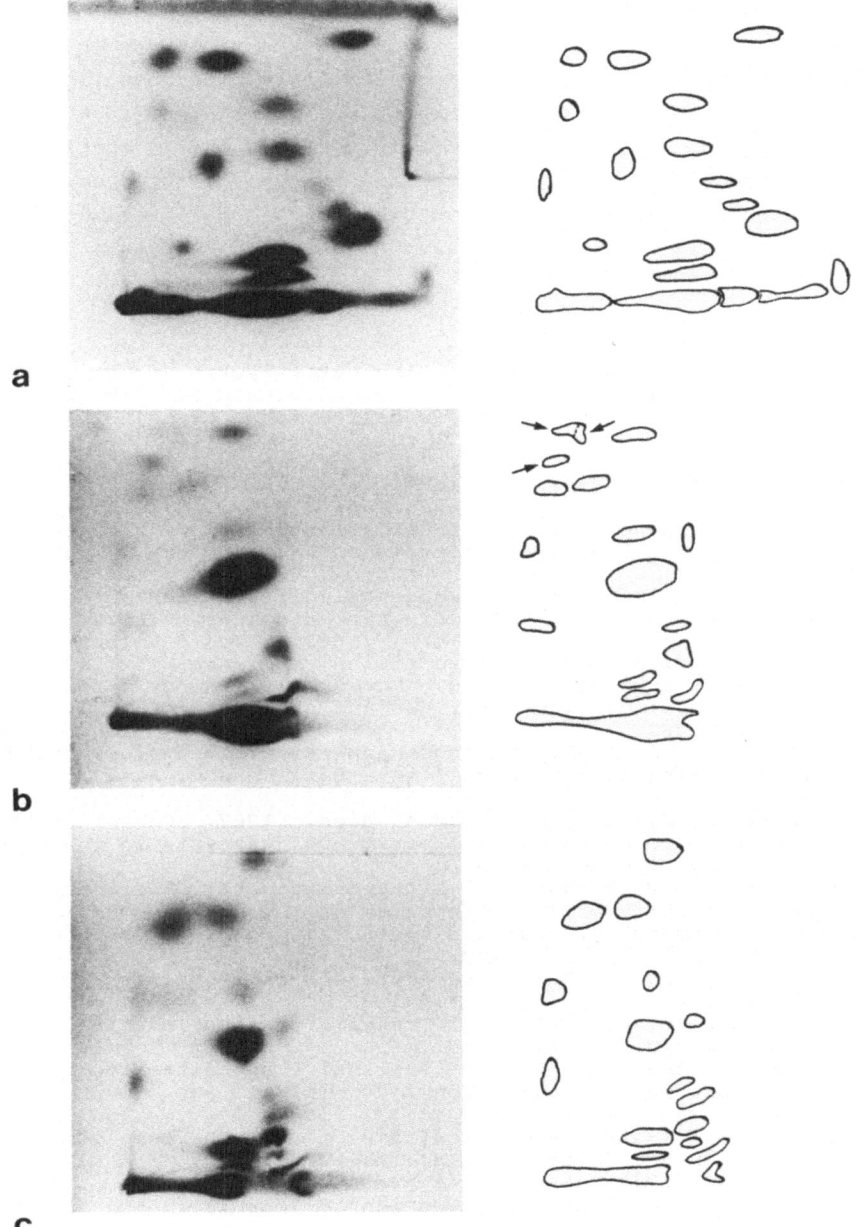

above mentioned results. This action of the lens is not limited to the chick but extends to the duck as well.

We conclude from these experiments that a substance is produced by chick and duck eyecups which causes the induction of a lens in overlying ectoderm. The substance is produced at a restricted point in the development of the eye, and is specific for lens induction. Tissues other than eyecups do not produce this substance. Immunological and biochemical characterization of this substance is still in progress.

REFERENCES

Beebe, D.C., D. E. Feagans, E. J. Blanchette-Mackie and M. E. Nau. 1979. Lens cell elongation without microtubules. The role of volume regulation in lens fiber cell morphogenesis. Anat. Rec., 193:3.

Brahma, S. K., J. Bours and W. J. van Doorenmaalen. 1971. Immunochemical studies of chick iris. Exp. Eye Res., 12:194.

Brahma, S. K. and W. J. van Doorenmaalen. 1971. Immunofluorescence studies of chick lens FISC and α-crystallin antigens during lens morphogenesis and development. Ophthal. Res., 2:344.

Coulombre, A. J. 1965. In "Organogenesis" (R. L. DeHaan and H. Ursprung, eds.), p219. Holt, Rinehart and Winston, N.Y.

Coulombre, A. J. and J. L. Coulombre. 1963. Lens development: fiber elongation and lens orientation. Science, 142:1489.

Cross, C. and C. Labouesse. 1969. Study of the dansylation reaction of amino acids, peptides and proteins. Eur. J. Biochem., 7:463.

Genis-Galvez, J. M. 1965. Quelques aspects de la differenciation et la determination du cristallin. Bull. Assoc. Anat., 49:642.

Genis-Galvez, J. M., L. Santos and A. Rios. 1967. Causal factors in corneal development: an experimental analysis in the chick embryo. Exp. Eye Res. 6:48.

Hay, E. D. and J. W. Dodson. 1973. Secretion of collagen by corneal epithelium. I. Morphology of the collagenous products produced by isolated epithelia grown on frozen-killed lens. J. Cell Biol., 57:190.

Jacobson, A. G. 1966. Inductive processes in embryonic development. Science, 152:25.

Karkinen-Jääskeläinen, M. 1978a. Transfilter lens induction in avian embryo. Diff., 12:31.

Karkinen-Jääskeläinen, M. 1978b. Permissive and directive interactions in lens induction. J. Embryol. exp. Morphol., 44:167.

McKeehan, M. S. 1951. Cytological aspects of embryonic lens induction in the chick. J. Exp. Zool., 117:31.

McKeehan, M. S. 1954. A quantitative study of self differentiation of transplanted lens primordia in the chick. J. Exp. Zool., 126:157.

McKeehan, M. S. 1958. Induction of portions of the chick lens without contact with the optic cup. Anat. Rec., 132:297.

Mizuno, T. 1970. Induction de cristallin in vitro chez le poulet en absence de la vesicule optique. C. R. Acad. Sci., Ser. D, 271:2190.

Mizuno, T. 1972. Lens differentiation in vitro in the absence of optic vesicle in the epiblast of chick blastoderm under the influence of skin dermis. J. Embryol. Exp. Morphol., 28:117.

Niu, M. C. and V. C. Twitty. 1953. The differentiation of gastrula-ectoderm in medium conditioned by axial mesoderm. Proc. Nat. Acad. Sci. U. S., 39:985.

Osserman, E. F. 1960. A modified technique of immunoelectro-phoresis facilitating the identification of specific protein arcs. J. Immunol. 84:93.

Sirlin, J. L. and S. K. Brahma. 1959. Studies on embryonic induction using radioactive tracers. The mobilization of protein components during induction of the lens. Develop. Biol. 1:234.

Toivonen, S., D. Tarin and L. Saxen. 1976. The transmission of morphogenetic signals from amphibian mesoderm to ectoderm in primary induction. Diff. 5:49.

van der Starre, H. 1972. Lens induction studies in vitro. Ophthal. Res., 3:20.

van der Starre, H. 1977. Biochemical investigation of lens induction in vitro. I. induction properties of the eyecup, ectodermal response. Acta Morph. Neerl.-Scand., 15:275.

van der Starre, H. 1978. Biochemical investigation of lens induction in vitro. II. Demonstration of the induction substance. Acta Morph. Neerl.-Scand., 16:109.

van Doorenmallen, W. J., S. K. Brahma and H. J. Hoenders. 1968. Preparation of specific antisera against chicken lens protein components. Experientia, 24:30.

Regulation of Lens Morphogenesis and Cataract Pathogenesis by Pituitary-dependent, Insulin-like Mitogens

Howard Rothstein, Basil Worgul, and Allan Weinsieder

"Then there is that little ball of cells which migrated from the skin and thrust itself into the mouth of the eye-stalk."
Sir Charles Sherrington, 1951

When we started the work to be described, we were not primarily concerned with communication or development in the strictest sense of these terms but with the stimulation of hyperplasia following mechanical wounding of the lens epithelium. Studies from one of our laboratories (Rothstein et al., 1964) as well as those of C.V. Harding, (Harding et al., 1959; Rothstein and Harding, 1962) and N. Rafferty (1963, 1965) showed that mechanical trauma results in the proliferation of lens epithelial cells in teleost, amphibian and mammalian organisms. Unknown to us at the time, similar findings had been reported in the late 19th. century by Th. Leber (1878) and later by Paul Knapp (1900 a, b). In 1882, the immortal Jacob Henle said, "Doch scheint es mir der Mühe werth, auf ein bisher unbeachtetes Organ die Aufmerksamkeit zu lenken, an welchem die Vermehrung der Zellen auf dem Wege der Karyokinese vor sich geht und an welchem sie vielleicht, bei richtiger Wahl des Zeitpunktes, mit grosserer Sicherheit verfolgt werden kann, als an manchen der bisher durchforschten Organe und Gewebe.". (It certainly seems worth the effort to turn attention to a previously neglected organ, where cell proliferation is achieved by karyokinesis which process could be followed more reliably by an appropriate choice of experimental intervals than in many organs and tissues that have, until now, been exhaustively investigated.) (Figure 1)

This statement appeared in the same year that the word mitosis was coined by Flemming of Kiel (1882). However, most students who concerned themselves with ophthalmic biology used Schleicher's

"karyokinesis" rather than Flemming's term. Whole mounts of lens
epithelium for histological purposes were introduced in 1952 by
Alma Howard (Howard, 1952) who was interested in the response of
the cells to X-rays. However, in Otto Becker's classic, Anatomie
Der Gesunden und Kranken Linse (1883) reference is made to
preparations of this type executed by Heinrich Müller as early
as February of 1856. Whole mounts allow one to view all of the
epithelial cells of the tissue at once; they are well suited for
auto-radiography, microspectrophotometry and microfluorimetry.
The first feature was well exploited by Harding (Harding et al.,
1959, 1960) and Von Sallmann (Von Sallmanan et al., 1962).

The modern studies demonstrated that DNA is synthesized by
the reacting cells before the burst of mitotic activity that is
unfailingly evoked by injury. Our own laboratory did a great deal
of work with amphibian organisms and we demonstrated that mitotic
activity is preceded by waves of RNA and DNA synthesis under both
in vivo and in vitro (organ culture) conditions. Each of the
synthetic episodes is a prerequisite for the appearance of
proliferative activity and if appropriate metabolic inhibitors are
used to block any of them, those that usually follow, do not in
fact, do so. These early studies have been reviewed (Rothstein,
1968; Harding et al., 1971). In addition to the changes already
mentioned, Briggs (Briggs et al., 1976a, b) indicated that
histones and nonhistone chromosomal proteins are also synthesized
during the cell cycle of lens epithelium. The histones are
manufactured during the period of DNA synthesis and if the
formation of DNA is blocked, so too, is that of the histone
fraction. Non-histone chromosomal proteins (NHCPs) seem to be
synthesized throughout the cell cycle. Their formation is not
confined to a particular stage. Later work performed in the
laboratory of John Reddan (Harding et al., 1971) at Oakland
University indicated that the changes we had found in Amphibia
also occur in mammals. A consistent finding has been that quite
soon after stimulation, either by injury or by explantation,
tritiated thymidine is incorporated to a greater extent in
experimental than in control samples. This seems to occur in

Figure 1   A reproduction of some of Jacob Henle's drawings of
dividing lens epithelial cells from newts and frogs. They were
rendered about a century ago. Henle was the first significant
student to advocate the use of lens epithelium as a system for the
study of cellular proliferation (J. Henle, 1882).

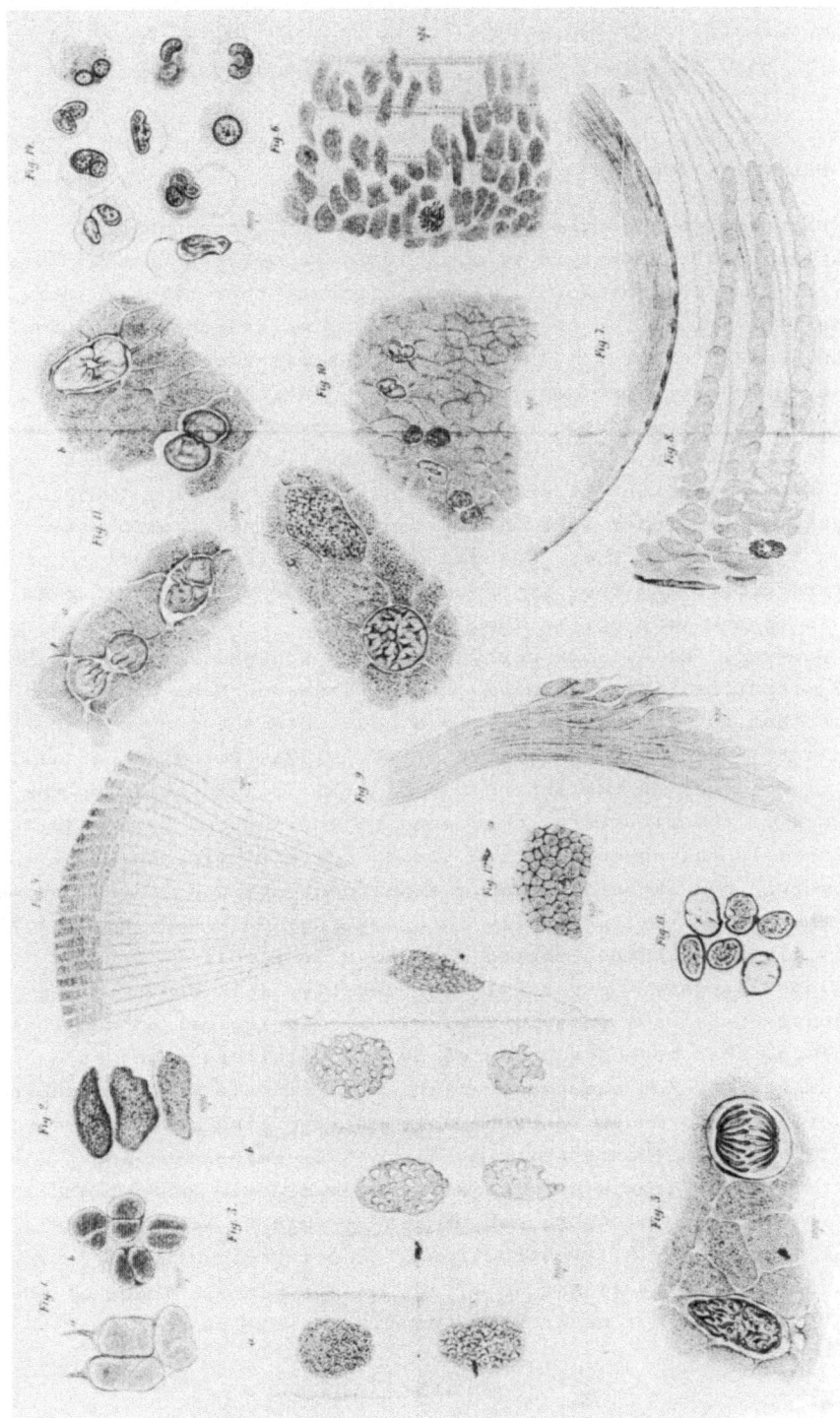

fact with any type of stimulation, and it occurs between one and
two hours after the stimulus has been applied (Weinsieder, 1973;
Worgul, 1974; Weinsieder et al., 1973; Worgul and Rothstein,
1974).

SEASONAL VARIATION

Experiments performed in our laboratory (Rosenbaum and
Rothstein, 1972; Rothstein et al., 1975) as well as that of
Sakharova and Golichenkov (1968) demonstrate that mitotic activity
in the lens epithelium of the frog undergoes seasonal variation.
Reddan (Reddan et al., 1975) has made a similar observation in a
marine teleost. Subsequent studies have indicated that this
variation may depend upon the action of hormones produced by the
anterior pituitary gland (Van Buskirk et al., 1975; Rothstein et
al., 1976; Wainwright et al., 1976; Rothstein, et al., 1979). The
highest mitotic index is found during the spring; the lowest
during the winter. These changes occur in nature as well as in
the laboratory, and they are found in the corneal epithelium as
well as in the lens epithelium.

However, following hypophysectomy by Hogben's method (Hogben,
1923) mitotic activity in Rana catesbeinana and Rana pipiens
lenses (but not corneas) comes to a halt within three to four
weeks after surgery (Van Buskirk et al., 1975; Rothstein et al.,
1976). So great is the dependence of cell proliferation on the
presence of the pituitary, that even injury induced hyperplasia is
prevented in the absence of the gland. Careful microspectro-
photometric and autoradiographic experiments by John Hayden showed
that the cells stall primarily in G_0/G_1 (Hayden, 1980; Wainwright
et al., 1978). In the leopard frog no S or M cell is found 3
weeks post surgery – a variable but usually small number of G_2
residents are always in evidence. It seemed logical to conclude,
therefore, that hormones produced by the pituitary might be
responsible for the seasonal variation in mitosis we had observed
earlier. The hormones found capable of reversing mitotic arrest
are: T3, T4, TSH (Van Buskirk et al., 1975; Weinsieder and
Roberts, 1980), growth hormone and frog prolactin (Wainwright et
al., 1976). Each of these substances is able to achieve this
effect independently. Destruction of the thyroid gland by
administration of ^{131}I has no effect on the mitotic index of the
lens epithelium, but under these conditions TSH is no longer

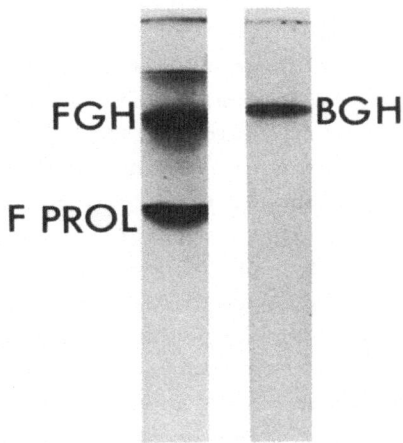

Figure 2   Separation of frog growth hormone and prolactin on a
polyacrylamide gel. The material is obtained from pooled anterior
pituitary glands of the southwestern frog Rana pipiens
berlandieri.   The faster migrating band (bottom) is thought to be
prolactin; the slower one, growth hormone. A gel containing
purified bovine growth hormone is shown at the right. FGH = Frog
Growth Hormone; F PROL = Frog Prolactin; BGH = Bovine Growth
Hormone (Wainwright et al., 1976).

mitogenic.   Growth hormone and prolactin have both been isolated
by gel electrophoresis, from frog anterior pituitary glands
(Figure 2).   Injection of these native hormones reinitiates
mitotic activity (Wainwright et al., 1976).   (It is noteworthy
that mammaliam prolactins do not have this capacity although
mammalian growth hormones do.) The authenticity of these
molecules has been verified by immunological techniques - these
indicate that the materials extracted actually originate in the
acidophillic cells of the pars distalis (Rothstein et al., 1979).

     Some years ago, it was observed that insulin stimulates
mitotic activity in rabbit and frog lenses cultured in medium 199
(Reddan et al., 1972).   The growth promoting effect of insulin has
been known for almost six decades, (Gey and Thalheimer, 1924) but
the striking response of organ cultured lenses to the hormone
called our attention to an interesting hypothesis that had been
proposed to explain the mechanism of action of growth hormone.

THE SOMATOMEDINS: INSULIN-LIKE GROWTH FACTORS

     Forty years ago, it was determined that longitudinal body
growth involves proliferation of chondrocytes in the epiphyseal
plates (Kibrick et al., 1941).   This activity ceases in
hypophysectomized animals and is restored by in vivo

administration of anterior pituitary extracts containing growth
hormone. Incorporation of tracer sulfate provides an index of the
growth of the cartilage. In the midfifties Salmon and Daughaday
(1957) observed that incorporation of $SO_4^=$ into rat costal
cartilage, maintained in vitro, does not respond to addition of GH
but increases promptly upon addition of serum from
hypophysectomized animals that have received the hormone in vivo.
Serum from non-injected, hypophysectomized rats has virtually no
activity. These workers proposed that the effects of GH are
actually mediated by a substance which they called "sulfation
factor". By 1972 sulfation factor was shown to have insulin-like
activity and to promote growth of some cells in tissue culture.
It was then called, by the more descriptive term, somatomedin
(Daughaday, et al., 1972); a molelcule that mediates the effects
of somatotropin.

Several peptides of the genre were isolated; these are
Somatomedins A (Fryklund et al., 1974), B (Yalow et al., 1975), C
(Van Wyk et al., 1975; Van Wyk et al., 1980), Non-suppressible
Insulin-like Activity (Rinderknecht and Humbel 1978a, 1978b -
[NSILA]) and Multiplication Stimulating Activity (Dulak et al.,
1973 [MSA]).

It was decided that in order to qualify for inclusion in such
a listing, a molecule should:

      1)   show synthetic dependency upon GH concentration.

      2)   enhance sulfate incorporation into cartilage.

      3)   exert insulin-like effects upon extraskeletal
      tissues.

Since somatomedin B did not fulfill criterion #2, it has been
dropped from this group of growth promoting peptides. Criterion
#3 notwithstanding, antiinsulin antibodies do not deprive these
substances of their ability to stimulate growth - hence the term
nonsuppressible insulin-like activity (NSILA). It has been
determined that NSILA is composed of two distinct molecules,
insulin growth factors I and II (Rinderknecht and Humbel, 1978a,
b). Weights of these insulin-like growth factors (ILGFs) range
from 7,000-10,000 Daltons and they are present in plasma. The
physiology of somatomedin-like molecules has been exhaustively
reviewed in recent years (Van Wyk et al., 1975, 1978, 1979, 1980;
Zapf et al., 1978).

We have observed that the hormones that were capable of
reversing the hypophysectomy effect in vivo were absolutely

ineffective in this regard under in vitro conditions (Wainwright et al., 1978). As mentioned, of the hormones tested, only insulin had this capacity, so we were driven to speculate that perhaps the effects of the anterior pituitary hormones tested are mediated in the same way as are the effects of growth hormone.

AN IN VIVO TEST FOR SOMATOMEDIN-C

Although we had, in the spring of 1978, ordered adult (300-700 g) bullfrogs, our commercial supplier actually sent us postmetamorphic animals that weighed 8-10 grams. Because of this circumstance we were placed in the fortunate position of being able to test pure somatomedin on an in vivo system. The absolute mitotic arrest produced by hypophysectomy in the frog lens was unique so far as we knew. One sticking point with the somatomedin hypothesis was that the substance had yet to be tested in vivo; it was simply too scarce for use in whole animal experiments. Through the generosity of Dr. J.J. Van Wyk of the University of North Carolina at Chapel Hill we were able to secure a 7µg sample of highly purified human somatomedin-C. It is a basic ILGF (pI 8.0-8.7) of approximately 7,500 Daltons which is separated from acid-ethanol extracts of Cohn fraction IV (Van Wyk, et al., 1980).

Seventy-five days after a postmetamorphic froglet had been hypophysectomized we began to inject somatomedin-C over an interval of an additional two weeks. The material was given each day and tritiated thymidine was injected every third day so that, in effect, a continuous label was achieved. As controls, hypophysectomized animals that were injected with physiological saline and isotope were used as were animals that had been hypophysectomized and injected with human growth hormone and ones that had not been operated upon at all, but had merely received physiological saline and isotope. The conditions and results of this experiment are shown in Figure 3. In the somatomedin injected organism, 24% of nuclei in the germinative zone incorporated tritiated thymidine. In the hypophysectomized control absolutely no such nuclei were found, whereas in the other two types of sample, 37-38% of the nuclei were engaged in DNA synthesis over the period studied. Thus somatomedin is able to reverse the hypophysectomy effect in vivo. Measurements of serum somatomedin by means of RIA revealed a molecule of this type in frogs. It is the lowest group of animals in which a

somatomedin-like molecule (SLM) has been found by means of RIA.
Investigations on fish have not given positive results although
these forms do have substances that promote sulfate incorporation
into cartilage (Shapiro and Pimstone, 1977). After hypophysectomy
the quantity of somatomedin-like material in the serum is reduced
by about half (Table 1); when growth hormone or T3 is given to a
hypophysectomized animal, the amount of SLM is elevated by an
approximately similar amount over the values found in intact
controls (Rothstein et al., 1980). The amount of somatomedin in
the aqueous humor represents about 7.5% of that found in the
serum, and hypophysectomy reduces this to about 0.4% of the amount
in controls. It must be admitted that measurements at this level
"push" the accuracy of the technique even though it is quite
sensitive. The assay is heterologous; one is measuring the
materials in frog serum with a rabbit anti-human SM-C antibody.
Nevertheless, we feel relatively safe in asserting that there is
an insulin-like molecule (SLM) in the serum of frogs whose
concentration is GH dependent. We believe it also responds to
frog prolactin and thyroidal substances.

Further, somatomedin-C triggers DNA synthesis in the
isolated lenses of hypophysectomized frogs in vitro as does
insulin (Hayden, 1980). Reddan (Reddan and Wilson, 1979) has
found that IGF-I and IGF-II as well as insulin have this effect on
the cultivated lenses of rabbit. So far, we have not found a
positive effect with IGF-I and II, even when derived from the same
batches observed to spark DNA synthesis in the rabbit lens. MSA,
on the other hand, is stimulatory. We have tested SM-C, MSA,

Figure 3   Autoradiograms of lens epithelial whole mounts (a) Lens
epithelium from a normal, intact, postmetamorphic bullfrog that
had been injected with Earle's balanced salt solution and
[^{3}H]thymidine, as described in the text. (b) A similar
preparation from a frog that had been hypophysectomized and had
received injections of isotope for 2 weeks. None of the cells
show evidence of DNA synthesis or mitosis. (c) Lens epithelium
from an animal that was hypophysectomized and later injected with
human growth hormone; labelling is evident. (d) Lens epithelium
from a hypophysectomized frog that received injections of purified
somatomedin-C starting on day 75 after hypophysectomy. This
animal received a total of ten 700-ng injections and was killed on
day 88 after the operation; note the mitotic figures and evidence
of incorporation of [^{3}H]thymidine. Cells manifesting these
phenomena were confined to the germinative zone in the tissue
shown in (d). The percentages of labelled cells are as follows:
(a) 37.7, (b) 0.0, (c) 39.7, and (d) 24.4 (Rothstein et al.,
1980).

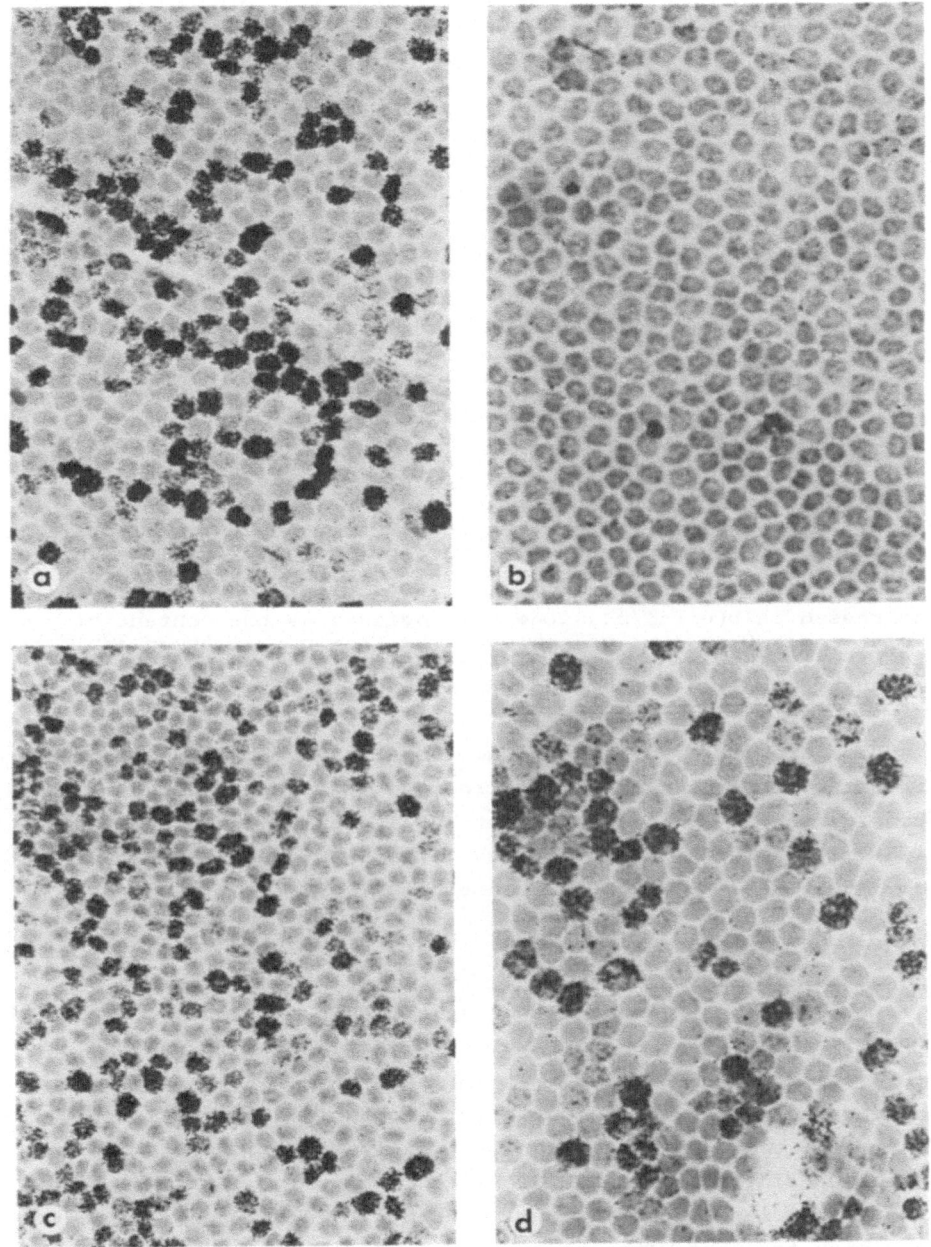

TABLE I

SOMATOMEDIN-C CONCENTRATION OF BULLFROG AND LEOPARD FROG SERA AND
AQUEOUS HUMOR (U/ml)

Experimental Status	Rana catesbeiana		Rana Pipiens
	Serum	Aqueous Humor	Serum
Normal (Intact Controls)	0.0475	0.0036	0.036
Hypophysectomized	0.0215	0.0002	0.024
Hypophysectomized, treated			
with 5 µg bGH/g body weight	0.079	-	-
Hypophysectomized, treated			
with 0.5 µg T_3/g body weight	-	-	0.043

The determinations were conducted by radioimmunoassay in the laboratory of J.J. Van Wyk. Serum was pooled from at least six animals. Hypophysectomy serum from bullfrogs was collected four weeks postoperatively. Replacement theory with bGH (bovine Growth Hormone) started at that time and was continued for four days before blood was collected. Hypophysectomy serum from leopard frogs was collected seven weeks postoperatively. Replacement therapy with T_3 commenced four weeks postoperatively and continued on alternate days until the time of sacrifice. Concentrations are expressed in U(nits)/ml. One U is defined as the content of somatomedin-C in 1 ml of pooled serum from normal human adult male subjects. The serum batch relative to which the reported measurements were determined was Ortho lot 1778-5.

insulin and IGF-I in mammals and frogs in vivo by injecting them into the vitreous chamber; we find that though the data must be considered preliminary - all but IGF-I seem to enhance mitosis. IGF-I is believed to be very much like SM-C so the results obviously need to be checked carefully.

THE ROLE OF THE LIVER

Most authorities hold that somatomedin is generated by liver (Phillips and Vassilopoulou-Sellin, 1980). Thus it was of interest when we observed that in frogs that have been hypophysectomized, histologically detectable damage (Figure 4) seemed to occur in that organ (Rothstein et al., in press - b). A series of experiments we did in collaboration with Drs. Edward Essner and Karni Frank at our institute, indicated a surfeit of glycogen together with steatosis. It had been reported by Snyder and Frye that the amount of carbohydrate increases in hypophysectomized frogs from 157 $\pm$ 18 mg/liver to 239 $\pm$ 30

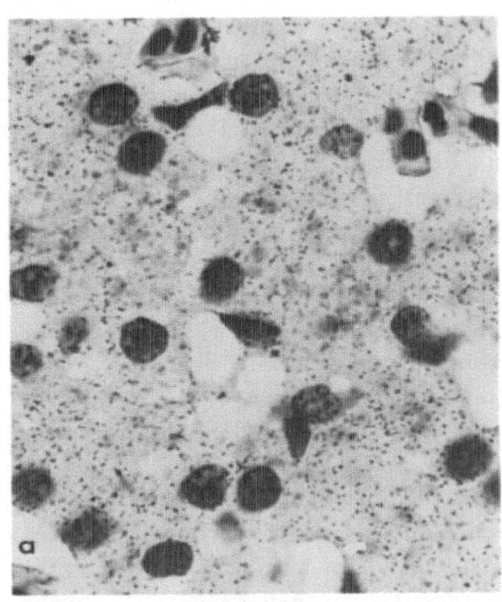

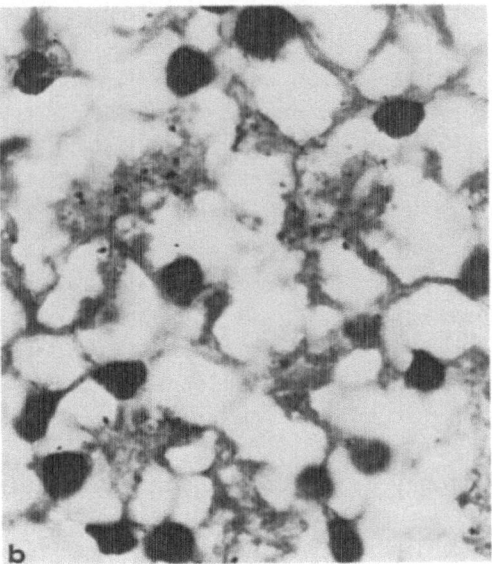

Figure 4 Autoradiograms of H and E stained liver sections from intact (a) and hypophysectomized (b) frogs. ^{125}I Human Growth Hormone (0.2 μCi/gram of bodyweight) was injected via the dorsal lymph sac. The animals were sacrified 20 minutes later. The tissue in (b) has undergone an obvious histopathological change but incorporation of the hormone is still evident in the peripheral bands of cytoplasm characteristic of parenchyma from hypophysectomized specimens (from Rothstein et al., in press - a).

mg/liver. Administration of GH to the latter class of animals
reduces the carbohydrate to 35 + 8.7 mg/liver while prolactin
reduces it to 141 + 28 mg/liver (Snyder and Frye, 1972). ·Others
have shown that the hepatocytes become less competent at protein
synthesis (Alford et al., 1976). Ribosome numbers decline and in
general the transcriptive machinery is not as efficient as in
control tissue.

Electron microscopy showed that glycogen changes from the α
to the β configuration. It was of interest to observe that
whereas growth hormone and T3 reverse the histologically
observable change in the liver, somatomedin-C does not. This
would be consistent with the proposal that the pituitary hormones
act upon the liver so as to generate ILGFs, which in turn regulate
the growth of the lens. In frogs as in mammals, growth hormone
binds to the hepatocyte (Weinseider and Rothstein, 1980). Studies
on blood chemistry show that hypophysectomy leads to decreases in
gamma glutamyl transpeptidase, triglycerides, and an increase in
the BUN. The operation also causes a very striking drop in blood
volume (Rothstein et al., in press - a). These changes are
corrected by administration of growth hormone. A drop also occurs
in the concentration of LDH which is not reversed by
administration of growth hormone, but is corrected by increased
feeding. It should be noted that some of the effects of
hypophysectomy are corrected by feeding with liver, but a
hypophysectomized frog will lose mitotic activity however well fed
it may be. Such is not necessarily the case for the mammal. Our
current view of control of lenticular mitotic activity is
represented in Figure 5 and Table II. We explain the seasonal
variation observed during the early seventies by the activity of
hypothalamic release and release inhibiting factors, then unknown
in amphibia, but which have since been identified by the
laboratories listed in the table.

MITOSIS, MIGRATION AND FIBERGENESIS

It has been stated by Hammar (1965) that "Mitosis in the lens
provides the force by which the epithelium cells are pushed toward
the bow area." [sic]   In fact, this was the consensus amongst
late 19th. and early 20th. century students of the lens (see for
example F.J. Von Becker, 1863). Movement of cells from the
germinative zone into the cortex has been followed by a number of

laboratories (Messier and Leblond, 1960; Brolin et al., 1961;
Hanna et al., 1963; Mikulicich and Young, 1963; Hammar, 1965;
Worgul and Merriam, 1976; Hayden and Rothstein, 1979). Most of
the studies have been done by means of tritium thymidine
autoradiography of sectioned materials. Just as there is nothing
in the literature to contradict the mode of fiber formation
suggested, neither has there been any irrefutably direct
substantiation of it. The ability to manipulate mitotic activity,
as discussed previously, permits such an approach. Hypophysectomy
can be instituted to reverse the effect. This sort of experiment
was conducted by John Hayden in H.R.'s laboratory in the late 70's
(Hayden and Rothstein, 1979).

The strategy was to administer labelled thymidine during the
last round of DNA synthesis that the cells in the germinative zone
would undergo following hypophysectomy. In this way he could
follow the fate of the tagged cells in the absence of sustained
division. Thymidine was introduced into hypophysectomized and
intact frogs one week after surgery. The animals were kept for as
long as 82 days after injection of isotope. In controls, the
distribution of labelled nuclei is relatively large ranging from
the precentral zone to the base of the meridional rows (the base
is the region in which the nuclei first begin to align in obvious
register -- see Figures 6 and 7). For example, in a control
sacrificed 56 days after injection of thymidine, 5-6 families of
nuclei could be recognized on the basis of grain counts (Figure
8). On the other hand, whole mounts from hypophysectomized
leopard frogs showed narrow bands of incorporating nuclei with but
one grain class (Figure 9). The labelled population divided no
more than once after administration of tritiated thymidine. The
cells seem not to have approached the meridional rows, whereas in
the control specimen, they are very near to them.

Is there then, a correlation between growth - as reflected in
labelling index - and migration of the cells into the bow?
Figures 10-13 imply that this is indeed the case. For the control
lenses, the correlation coefficient between percentage of cells
labelled and movement toward the base of the rows is 0.86. The
correlation coefficient between position of such cells and time is
0.87. When, however, one computes the correlation coefficient as
a function of time for animals that have been hypophysectomized,
it turns out to be 0.02. It would then seem that growth is a more
certain indicator of migration than is time. In fact, it may be

124

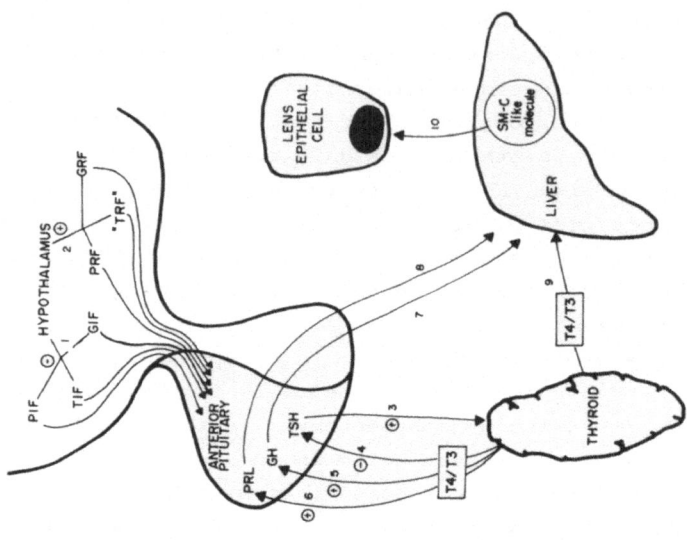

Figure 5   Control of Mitosis in the lens epithelium of
the frog. The hypothalamus secretes releasing and
release-inhibiting peptides that influence secretion of
prolactin, TSH and growth hormone from the pituitary.
These hypophysiotropic factors are under the control of
environmental (e.g., seasonal effects) and internal
influences. Enhanced production of TSH leads to
increases in T4/T3 which can act directly upon the liver
to cause formation of somatomedin-like substances which
in turn augment cell division in the lens. In addition,
the thyroidal material causes production of the mRNA for
growth hormone in the acidophilic cells of the anterior
pituitary. It also causes secretion of growth hormone
and prolactin. Further, growth hormone and prolactin
cause generation of mitogenic peptides resembling
somatomedin-C by hepatocytes. The mitogen(s) trigger
movement of lens epithelial cells from the G_0 into the
G_1 segment of the cell cycle. The numbers over the
arrows refer to some studies in which the relevant
evidence has been presented. These numbers with
corresponding references are listed in Table II.
(+) Stimulatory; (-) Inhibitory
(from Rothstein et al., in press - a)

TABLE II: STUDIES CORROBORATING THE MODEL FOR GROWTH CONTROL DEPICTED IN FIGURE 5

Diagram Number	Event	References	Diagram Number	Event	References
1	Prolactin and thyrotropic inhibitory activity in hypothalamus of Rana temporaria	Kuhn and Engelen, 1976	7	GH stimulates release of somatomedin for liver (mammals)	Schalch, et al., 1979
1	Growth hormone inhibitory factor in hypothalami of anura	King, et al., 1979 Vandesande and Dierickx, 1980	7	GH increases levels of somatomedin in hypopituitary mice	Holder and Wallis, 1977
2	Thyrotropic hormone-releasing hormone in hypothalamus of Xenopus laevis	Goos, 1978	7	Mammalian and frog GH stimulate cell division in the frog lens	Van Buskirk, et al., 1975; Wainwright, et al., 1976
2	Prolactin-releasing factor in hypothalamus of Rana temporaria	Kuhn and Engelen, 1976	7	bGH elevates level of SLM in bullfrogs	Rothstein, et al., 1980
2	GH and Prolactin-releasing activity in hypothalamus of Xenopus laevis	Hall and Chadwick, 1979	8	Prolactin increases levels of somatomedin in mammals	Holder and Wallis, 1977
3	TSH stimulates release of thyroxin in Rana catesbeiana	MacKenzie, et al., 1978	8	Frog prolactin stimulates cell division in lenses of intact and hypophysectomized frogs	Wainwright, et al., 1976
3	TSH stimulates cell division in lenses of Rana pipiens	Van Buskirk, et al., 1975	9	T4/T3 stimulates cell division in lenses of intact and hypophysectomized frogs	Van Buskirk, et al., 1975; Weinsieder and Roberts, 1980
4	Removal of thyroid gland leads to increase in size of TSH producing cells in newts	Compher and Dent, 1970	9	T4 stimulates increases in somatomedin in hypophysectomized rats	Gaspard, et al., 1978
5	T4/T3 stimulates increase in GH mRNA in pituitary cells (mammals)	Martial, et al., Samuels and Shapiro, 1976	9	T4 increases levels of somatomedin (mammals)	Holder and Wallis 1977
6	T4/T3 stimulates prolactin secretion (mammals)	Ensor, 1978	9	T4/T3 binds to amphibian hepatocytes	Kistler, et al., 1975; Galton, 1980
7	hGH binds to frog liver	Weinsieder and Rothstein, 1980	9	The number of liver binding sites for hGH is reduced in the hypothyroid rat	Duran and Garcia, et al., 1979
			9	Decrease in T4 leads to a decrease in somatomedin (mammals)	Burstein, et al., 1979
			9	T3 increases levels of SLM in Rana pipiens	Rothstein, et al., 1980

that some disagreements in the literature are based on the fact
that investigators have worked with systems in which different
amounts of mitotic activity were in progress. It should be
stressed that one cannot distinguish, in this work, between free
cell locomotion and mass movement but only transposition of
labelled cells over time.

The results to date suggest that migration halts in the
absence of cell division. Does this affect the deposition of lens
fibers, that is morphogenesis? If fiber formation were to
continue in the absence of cell renewal, one would, a priori,
anticipate that the cell population density from the base of the
MR to the bow would become progressively sparser. In fact, no
such observation is made. The average population density in
intact Rana pipiens is 44 ± 1.22 SD while that in
hypophysectomized organisms is 48.0 ± 5.55 SD. In addition the

Figure 6   Drawing of the frog (Rana pipiens) lens. The
epithelial cells cover the anterior surface of the organ (top) and
are themselves covered by an elastic capsule. These cells may be
removed and a whole mount made as shown in the smaller drawings
(bottom left and right). The technique is feasible because the
cells adhere firmly to the capsule after appropriate fixation.
Though the lens epithelium consists of one cell type these cells
have differing proliferative capabilities and they are arranged in
a number of zones as indicated in the drawing. (CZ = Central
zone; PCZ = Precentral zone; GZ = Germinative zone; TZ =
Transitional zone; MR Meridional rows). Mitotic activity is
generally confined to the GZ. Lens fibers are formed from
epithelial cells that migrate from the GZ into the cortex of the
lens. Eventually the nuclei are lost and the fibers elongate
toward the anterior and posterior poles (from Hayden and
Rothstein, 1979).

Figure 7   These drawings show enlargements of regions A and B
from Figure 6. A is a portion of a whole mount and B is a portion
of a sectioned lens. The most important zones from the point of
view of the present report are presented so they can be seen from
both vantage points simultaneously. Particularly noteworthy are
the areas to which mitosis is confined and the lining up of the
cells into meridional rows (MR). Whole mounts are clearly
favorable for evaluation of mitotic activity while sections are
indispensable for following fibergenesis. Unlike other
regenerating tissues the lens does not shed any cells or fibers.
The latter are continually pushed deeper into the organ - which
thus contains an ever present record of its past history. A = Arc
of the lens bow; SPA = Superficial post arcuate zone; DPA = Deep
post arcuate zone; RFD = Region of fiber denucleation (from Hayden
and Rothstein, 1979).

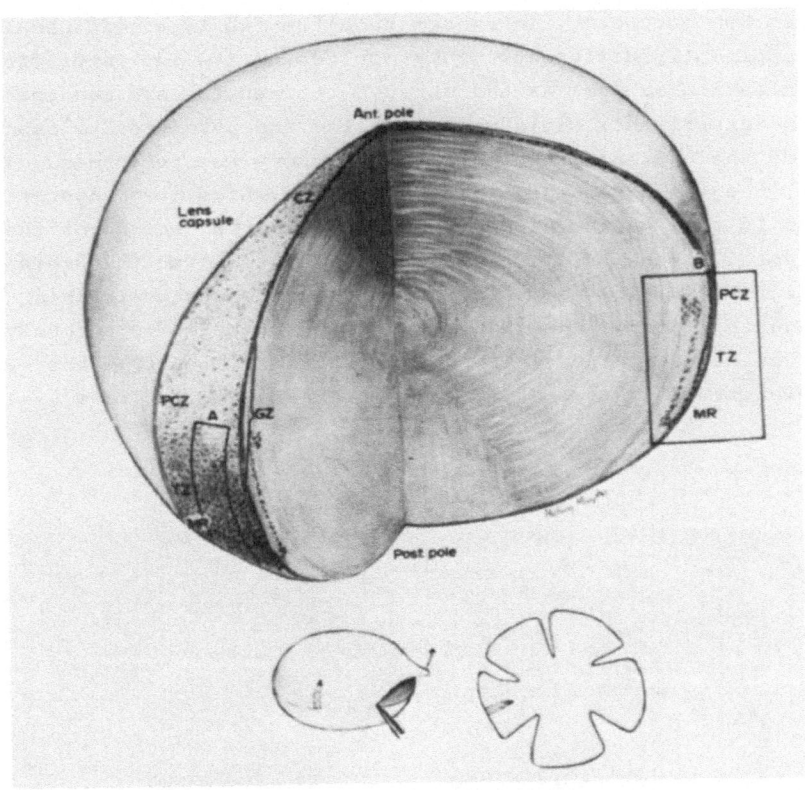

6

7

overal bow morphology in lenses from the two types of animals are not apparently different. This applies to the region of fiber denucleation as well as the stretch between the arc and the RFD. In one experiment, cells were labelled and permitted to reach the arc of the bow, after which hypophysectomy was performed. In this case, the cells remained where they were while in nonoperated controls that had been permitted to reach the same point they proceeded to the RFD. Of course one must realize that this type of migration involves perigrinations of the nucleus within the elongating fiber, whereas the migration spoken of previously is that of cells in the centrifugal direction (i.e. from the germinative zone toward the base of the meridional rows). In the

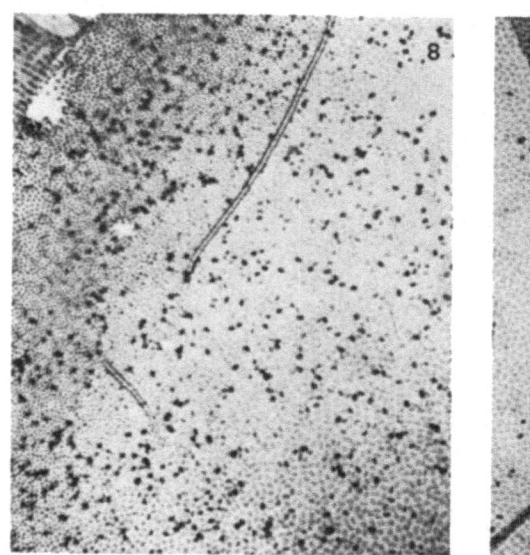 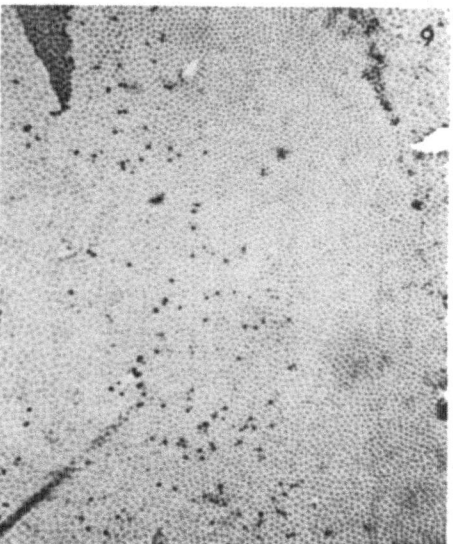

Figures 8 & 9   Autoradiograms of whole mounts from intact (8) and hypophysectomized (9) frogs.   The former animal was given bGH. The animal was injected with ^{3}H-thymidine and sacrificed 56 days later.   The breadth of the zone of incorporation is relatively great and one can discern six families of nuclear grain densities (not visible at this magnification).   In (9) the animal was hypophysectomized and injected with isotope after one week.   Then, 82 days after the injection the animal was killed.   Despite the additional 26 days allowed in this instance the band of incorporation is narrow and only one class of nuclear grain densities was seen.   These observations suggest that both division and migration have been arrested in the hypophysioprivic frog (from Hayden and Rothstein, 1979).

absence of mitotic activity movement of both types comes to a
halt. We cannot be certain, however, that fibergenesis itself
stops.

In one group of experiments we labelled the cortical fibers
by means of intravitreal injection of high specific activity
tritiated leucine. The amount of incorporation into the lenses of
hypophysectomized animals seems somewhat higher than in those of
the controls. Without rendering an interpretation of this
observation, it may be said that the subsequent distribution of
isotope (over a period of two months) is similar in
hypophysectomized and intact frogs, but it must be borne in mind
that the centripetal dispersion of tritium may be due to diffusion
as well as to the laying down of fresh fibers. Isolation of
proteins is currently being conducted by Dr. Sheldon Gordon with
the help of Dr. Mihir Bagchi.

G_O/G_1 ARREST OF CELL PROLIFERATION AND RADIATION CATARACT

"Die Thatsache, dass jede Kataraktbildung von einer Wucherung
der intracapsulären Zellen begleitet ist, eröffnet neue
Gesichtspunkte, deren Consequenzen noch nicht abzusehen sind."
(The fact that every cataract formation is accompanied by
proliferation of the intracapsular cells, opens new viewpoints
whose consequences cannot yet be foreseen." (Otto Becker, 1883)

It is common knowledge that ocular irradiation causes
posterior subcapsular cataracts. It has been proposed by many
authors that this phenomenon depends upon some form of injury to
the epithelium (Grzedzielski, 1935; Poppe, 1942a, b; Cogan, 1952;
Cogan et al., 1952; Worgul and Rothstein, 1974). It is further
held that continuing mitotic activity is necessary for
opacification to occur, because in frogs with a normally low
mitotic index or in which this effect is achieved experimentally,
radiation cataracts mature less readily than in the normals
(Worgul and Rothstein, 1974). Mitotic activity has also been
judged to be the basis for the inverse relationship between
cataract susceptibility and age in mammals (Worgul, 1974; Worgul
and Rothstein, 1974).

The postmetamorphic froglet has been a boon for our studies
of cataractogenesis. Its lens grows quickly, and as shown, this
growth depends upon anterior pituitary outflow. In an adult
leopard frog, it takes from 4-5 months for germinative zone cells

130

labelled with tritiated thymidine to reach the arc of the bow
(Hayden, 1980) whereas it requires only 39 days for such a cell in
a postmetamorphic animal to travel to the region of fiber
denucleation. In an intact control animal irradiation of the eye
leads to a disruption of the meridional rows (MR) and then to a
posterior subcapsular cataract. Rabl (1900) named and laid stress
on the MR in lens organization. Recently, Streeten and Eshaghian
have reemphasized the role of posterior epithelialization in the
generation of posterior subcapsular cataracts in the human lens
(Streeten and Eshaghian, 1978; Eshaghian and Streeten, 1980). The
opacities that form at the posterior pole are believed to result
from inappropriate migration of cells in the arc. They move
posteriorly instead of toward the RFD. When they reach the back
of the lens, where no cell should be, in normal adults, their

Figure 10 Plot of cellular migration against percentage of
labelled cells in intact animals. The lens epithelium was tagged
with 3H-thymidine by injection of the isotope into the dorsal
lymph sac. The animals were sacrificed at a number of subsequent
periods (from 0 to 8 weeks). The percentages of cells labelled
with the isotope were calculated from autoradiograms of whole
mounts and sections. On the ordinate are indicated various
regions of the lens as shown in Figures 6 and 7. The distance of
migration of the cells was measured using the base of the
meridional rows as a reference point. (The distance of GZ cells
from the base of the rows may, however, vary from 300 to 700 µm.)
The points represent the location of the labelled cells as they
progress toward the meridional rows. For each preparation an
average of 40 measurements was taken to obtain the distance of
migration of the labelled cells from the GZ toward the MR. The
correlation coefficient for this plot is 0.86 (from Hayden and
Rothstein, 1979).

Figure 11 This plot is similar to the one shown in Figure 10 but
contains data points for hypophysectomized animals and intact
animals that received bGH as well as untreated intact animals.
The hypophysectomized animals never had more than 6% of the cells
labelled and they have not moved closer than 300 µm to the base of
the MR. In the bGH treated frogs, cells are already in the lens
bow (from Hayden and Rothstein, 1979).

Figure 12 Plot of position (ordinate) of labelled cells against
time (abscissa) in the lens epithelium of intact frogs. This plot
was developed like those shown in Figures 10 and 11. However, in
this case, position was noted as a function not of labelling index
but of time (6 h to 8 weeks). The correlation coefficient is 0.87
(from Hayden and Rothstein, 1979).

Figure 13 Plot of position (ordinate) of labelled cells against
time (abscissa) in the lens epithelium of hypophysectomized frogs.
The correlation coefficient for these parameters is 0.02. Thus,
migration is not a demonstrable function of time under the
conditions of the experiment described (from Hayden and Rothstein,
1979).

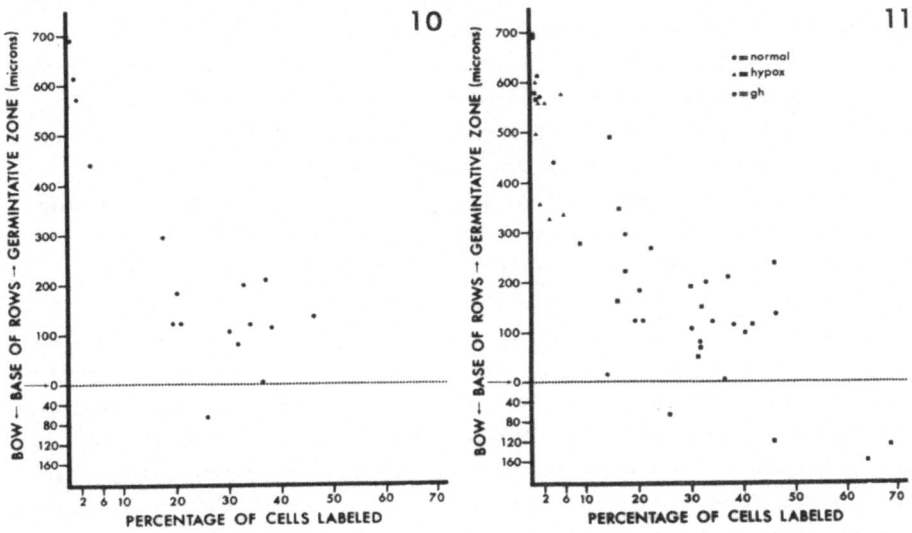

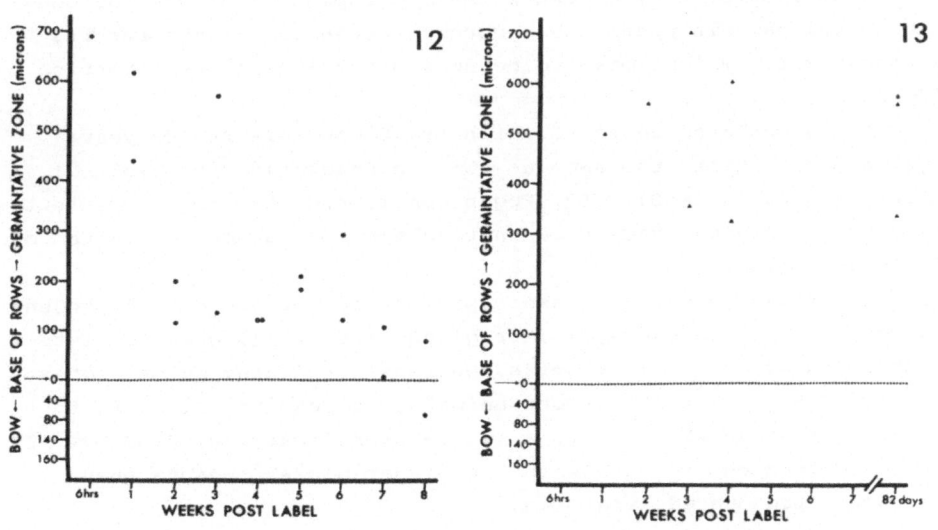

differentiation becomes decidedly abnormal. This was noted by Grzedzielski (1935) and Poppe (1942a, b) in the late thirties and early forties. Some years ago, Dr. David Cogan (Cogan, 1952; Cogan et al., 1952) obtained 26 human lenses that had been exposed to either X or gamma rays - some were from atomic bomb victims. Based on studies of these samples, Dr. Cogan says "...and these damaged cells reach the equator. Here normally the cells differentiate into lens fibers, as you know, elongating, sending fibers out forward and backward, and their nuclei ultimately disappear. In a lens that has been damaged those cells that cannot differentiate into lens fibers pile up at the equator.

"After the cells have piled up some migrate. I don't know whether it is active migration or passive, but I suspect it may be passive. They migrate beneath the only potential space; that is, between the capsule and the cortex. They migrate beneath the posterior capsule and come to lodge chiefly, in human beings, in this central area, right at the posterior pole of the lens....What is there about the posterior pole that makes them pile up in that region? ...That is the place of least mechanical resistance to the piling up of the cells. ...After these cells lodge there they appear to undergo an abortive differentiation into lens fibers...instead of increasing their cytoplasm and spreading out into lens fibers, they increase their cytoplasm and form balloons; the nuclei may disappear and a round globule or balloon results - a viable cell, which looks to be an aborted lens fiber formation (Cogan, 1952)."

The vacuolated cells to which Dr. Cogan referred 29 years ago, are of course, the same as those characterized by Wedl 121 years ago (Wedl, 1860). Dr. Cogan considered that they represent a cohort so placed that it can not adhere to its normal program of differentiation.

In frogs the cataractogenic process is age and dose dependent (Worgul et al., in press). Thus given 25 Gy, 8-12 gram froglets develop cataracts in four weeks, whereas 15-20 gram animals do so in nine weeks. For the larger animals, it requires 11 weeks to reach this stage with 10 Gy, a little over 7 weeks with 50 Gy. It is not convenient to work with the higher doses, because they produce a pronounced keratitis.

What of the effects of mitostasis? What occurs following hypophysectomy? In Rana pipiens and in bull froglets the MR retain normal alignment after exposure to X-rays; cataracts do not

appear (Hayden et al., 1980). Are these specific effects of
arrested growth or are they present because of the general
influence of hypophysectomy? To examine this question, we chose
the rat whose lens growth remains unchanged after anterior
pituitary ablation.

Frogs and rats received 6 Gy per minute for a total of 25 and
12 Gy respectively. The amphibia developed no cataracts over the
17 weeks observation period (Figure 14). Intact controls had at
least 2 + opacities. The morphology of the cataract in the
froglets is very characteristic, first appearing as striae
festooned with vacuoles approximating the meridians on the
posterior surface of the lens. Later, larger vacuoles and/or
opacities, similar to those seen in mammals develop. No such
configurations are seen in the hypophysioprivic froglet although
one occasionally sees punctate opacities in unirradiated, intact
as well as in unirradiated hypophysectomized animals.

In striking contrast, irradiated rats develop cataracts after
the surgery (Figure 15), though there may be a slight difference

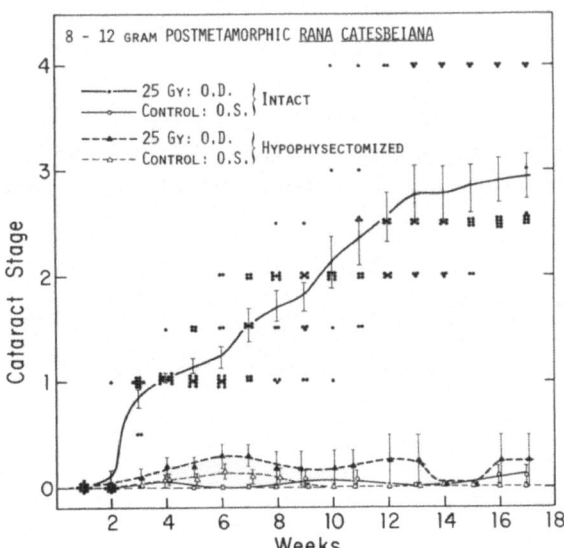

Figure 14 Cataract development in intact and hypophysectomized
froglets. The data for the intact animals are presented in the
form of a scatter plot. Each point represents an irradiated lens.
The line traverses the means and the error bar represents the SEM
of each mean. The remaining data are plotted as the means of at
least 6 lenses ± SEM.

134

in the chronology of their maturation. Therefore, extirpation of
the pars distalis stops the cataractogenic process in frogs but
not rats. Why? We believe it is because mitosis is stopped in
the amphibia but not in the rodents. Evidence has been adduced to
suggest that migration of lens epithelial cells is a mitosis
dependent phenomenon. In its absence the cells affected by
radiation are unable, therefore, to reach the back of the lens.
Because removal of the pituitary does not significantly hamper
proliferation in rat lenses the travels of the cells which foster
opacification would not be impeded either. We predict that if a
way could be found to achieve G_0 arrest in the mammalian lens, it
would also preclude the development of posterior subcapsular
cataracts.

SPECIFICITY

That insulin-like growth factors play a prominent role in
mammals is accepted. That these molecules are involved in growth
regulation in lenses of such organisms is likely. For example, it
has been observed that insulin, IGF I and II stimulate the
cultured rabbit epithelium to divide (Reddan and Wilson, 1979).
We have observed that MSA and insulin elicit division after
intravitreal injection into rabbits and rats, but hypophysectomy
does not change the MI of the rat lens (Figure 16). If ILGFs
control proliferation in rat lenses, and if hypophysectomy does

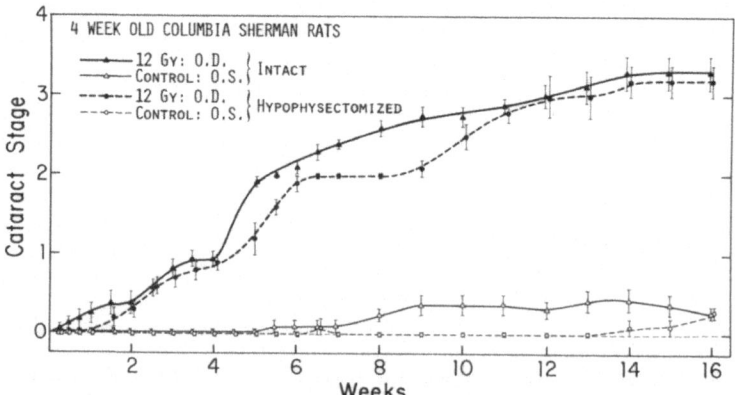

Figure 15 The development of radiation cataracts in the eyes of
irradiated, intact and hypophysectomized rats. Each point is the
mean of 8 eyes ± SEM.

not interfere with such proliferation one may infer that in rats
the procedure does not lower the concentration of pertinent growth
factors sufficiently to produce G_0 arrest.

It has become abundantly clear that nutrition exerts a major
influence on the generation of somatomedins in mammals (Phillips
and Vassilopoulou-Sellin, 1980). In kwashiorkor, the serum
somatomedin-C level declines to very low levels even in the face
of normal or elevated growth hormone concentrations. Conversely,
Phillips and Vassilopoulou-Sellin say that "Obese children and
children with hyperphagia after operations for hypothalamic tumors
(such as craniopharyngiomas) may have good growth with normal
somatomedin activity even though growth hormone is low or absent,
in these children, good nutrition and normal or elevated insulin
appear to sustain somatomedin activity in the presence of low
growth hormone" (Phillips and Vassilopoulou-Sellin, 1980). In
subsequent experimental studies Vassilopoulou-Sellin et al.,
(1980) observed that livers from well-fed rats produce relatively
large amounts of somatomedin upon perfusion with GH, but that
livers from fasted animals could not produce much of the growth
factor even when supplied with GH. Insulin also seems to play an
important role in regulating the effects of GH and stimulates
growth on its own. Perfusates from fasted and diabetic rats,
decreased cartilage stimulation by serum growth factors between 45

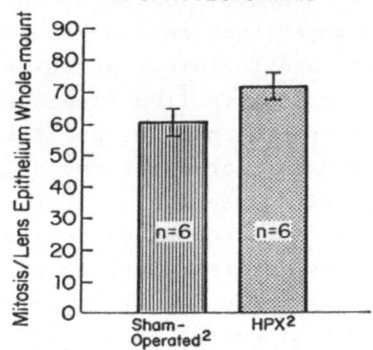

MITOTIC ACTIVITY IN LENS EPITHELIUM
FIVE WEEKS AFTER WEANLING RATS WERE
HYPOPHYSECTOMIZED[1]

[1]Mean weights of rats at time of sacrifice: Controls,
313g; Hpx, 80g.

[2]Representative whole-mount preparations of lenses
derived from 4 animals in each group were evaluated.

Figure 16   Effect of hypophysectomy on lens epithelial mitosis of
weanling rats on standard diet.

and 61% when added to media containing normal serum, suggesting, that these perfusates contain somatomedin inhibitors too. The point of these findings - and there are other suggestive ones which space does not permit us to discuss - is that in the mammals liver generated insulin-like mitogens may be very much dependent upon nutritional status. We have done some preliminary studies with rats and they further corroborate this idea.

Normal Columbia Sherman rats weighing 150-250 grams, generally have 75-120 mitotic figures. This varies diurnally as Von Sallmann and Grimes (1966) showed. When such rats are hypophysectomized and starved for three days, the mean mitotic activity in the lenses drops to 15 or 16; (ca 90% inhibition). The hypophysectomized animal which is fed has about the same amount of mitosis as a normal one. An intact animal that is starved over this period of time shows a drop of about 40% in mitotic activity. If hypophysectomized animals are not fed protein for a month, (which is about all that they can survive), the mitotic index drops by 85-90%. Hypophysectomized animals with complete diets again have about the same amount of mitosis as do intact, fed animals. The intact animals on protein deficient diets witness a 50% reduction in mitosis. The effect takes some three months to appear. The data are preliminary but consistent with the possibility that somatomedin like molecules regulate proliferation in the rat lens. A pilot study with Rhesus monkeys (done in collaboration with Dr. R. Norman of the Oregon Primate Center) showed that hypophysectomy causes reduction in plasma somatomedin-C (Table III) and incidence of ^{3}H-thymidine incorporating nuclei in lens epithelium (Table IV). More work needs to be done in this area. Species specificity might reside in the particular dependencies for somatomedin generation in given organisms not in the actual regulation of lens cell proliferation by these molecules. What of tissue specificity?

Neither hypophysectomy nor introduction of growth factors or hormones into frogs has any influence upon the mitotic activity which occurs as the result of corneal endothelial wounding. Frog corneal epithelium is similarly insensitive to these procedures, but the tissue specificity is certainly not complete. Cartilage obviously does depend upon ILGFs, and a recent report asserts that the growth of arterial smooth muscle cells may be so regulated (Weinstein, et al., 1981). In addition to somatomedins

TABLE III

HETEROLOGOUS RADIOIMMUNOASSAY FOR SOMATOMEDIN IN RHESUS PLASMA[1,2]

Animal	Somatomedin (U/ml)	
	pre-operative	post-operative
Sham-Operated		
#9779	3.05	-
#9797	4.40	-
#9930	3.05	3.41
#9931	2.37	3.24
mean	3.21	3.35
Hypophysectomized		
#9988	2.73	0.33
#9991	5.42	0.27
#9993	3.72	<0.21
#10002	3.36	0.30
mean	3.81	0.35

[1]Determined by Dr. J.J. Van Wyk (Univ. North Carolina, Chapel Hill) with antiserum against human SM-C raised in rabbits (Furlanetto et al., J. Clin. Endocrinol. 39, 283, 1974); EDTA-treated plasma was acidified prior to assay, as suggested by Dr. L. Underwood.

[2]A Unit is defined as the content of somatomedin-C in 1 ml of pooled serum from normal human adult males.

TABLE IV

EFFECT OF HYPOPHYSECTOMY
ON THE NUMBER OF LENS EPITHELIAL
CELL NUCLEI LABELLED BY TRITIATED THYMIDINE[1,2]

	Evaluation based on:		Labelled cells:[3]	
	preparations	# of animals	per grid	per 10³
Sham-Operated	8	4	4.6	2.7
Hypophysectomized	4	3	2.8	1.6
Effect of Hypophysectomy			39% reduction in number of cells labelled	

[1]Thymidine - ^{3}H(0.67 uCi/gbw; 60 Ci/mM; Schwarz-Mann) given I.V. 24, 16 and 8 hours prior to sacrifice.

[2]Autoradiograms (NTB-2, Kodak) of Feulgen-stained whole-mount preparations of lens epithelium exposed 17 days.

[3]All labelled cells falling within the area of one eyepiece net micrometer, aligned at a standard distance from the base of the meridional rows, were scored; a sampled area at 25 x contained approximately 1700 cells; minimally, 6 areas per preparation were evaluated.

transferrin, insulin, thrombin, platelet derived growth factor, epidermal growth factor, hydrocortisone, and other materials were found necessary to promote maximal proliferation. One is inclined to conclude that in many situations a multiciplicity of factors and conditions may be important in growth regulation. Because of its unique qualities the lens has provided us with unique experimental opportunities to define some of these factors and conditions.

REFERENCES

Alford, F.P., M. Millea and G.M. Reaven. 1976. Effect of food deprivation and hypophysectomy on in vitro protein synthesis by membrane-bound and free hepatic ribosomes. Horm. Metab. Res. 8:118-122.

Becker, O. 1883. Zur Anatomie der Gesunden und Kranken Linse. Wiesbaden. Verlag von J.F. Bergmann.

Briggs, R., H. Rothstein and N.R. Wainwright. 1976a. Cell cycle variations in chromosomal proteins of the lens epithelium. Exptl. Cell Res. 99:95-105.

Briggs, R., N.R. Wainwright and H. Rothstein. 1976b. On the chromosomal proteins of the lens epithelium. Documenta Ophthalmologica 8:57-65.

Brolin, S.E., H. Diderholm and H. Hammar. 1961. An autoradiographic study on cell migration in the eye lens epithelium. Acta Soc. Med. 66:43-48.

Burstein, P.J., B. Draznin C.J. Johnson and D.S. Schalch. 1979. The effect of hypothyroidism on growth, serum growth hormone, the growth hormone - dependent somatomedin, insulin-like factor, and its carrier protein in rats. Endocrinology 104:1107-1111.

Cogan, D. 1952. Pathogenesis of radiation cataracts. U.S. Atomic Energy Commission, National Research Council, Washington, D.C., Proceedings of January 28, pp. 1-8.

Cogan, D., D.D. Donaldson and A.B. Reese. 1952. Clinical and pathological characteristics of radiation cataract. Arch. Ophthalmol. 47:55-70.

Compher, M.R. and J.N. Dent. 1970. Responses to thiourea and to surgical thyroidectomy by the autotransplanted pituitary gland in the read spotted newt. Gen. Comp. Endo. 14:141-147.

Daughaday, W.H., K. Hall, M.S. Raben, W.D. Salmon, J.L. Van Den Brande and J.J. Van Wyk. 1972. Somatomedin: Proposed designation for sulphation factor. Nature 235:107.

Dulak, N.C. and H.M. Temin. 1973. A partially purified polypeptide fraction from rat liver cell conditioned medium with multiplication-stimulating activity for embryo fibroblasts. J. Cell Physiol. 81:153-160.

Duran-Garcia, S., J. Gomez-Nieto, M. Fouchereau-Peron, V.F. Padron, M.J. Obregon, G. Borreale De Escobar and F. Escobar Del Ray. 1979. Effects of thyroid hormones on liver binding sites for human growth hormone, as studied in the rat. Clin. Endocrinology 11:275-289.

Ensor, D.M. 1978. Comparative Endocrinology of Prolactin, Chapman and Hall, London.

Eshaghian, J. and B.W. Streeten. 1980. Human posterior subcapsular cataract: an ultrastructural study of posteriorly migrating cells. Arch. Ophthalmol. 98:134-143.

Flemming, W. 1882. Zellsubstanz, Kern-und Zelltheilung. Leipzig - see page 376.

Fryklund, L., K. Uthne and H. Sievertsson. 1974. Identification of two somatomedin A active polypeptides and in vivo effects of somatomedin A concentrate. Biochem. Biophys. Res. Commun. 61:957-62.

Galton, V.A. 1980. Binding of thyroid hormones in vivo by hepatic nuclei of Rana catesbeiana tadpoles. Endocrinology 106:859-866.

Gaspard, T., R. Wondergem, M. Hamadzie and H.M. Klitgaard. 1978. Serum somatomedin stimulation in throxine-treated hypophysectomized rats. Endocrinology 102:606-611.

Gey, G. and W. Thalheimer. 1924. Observations of the effects of insulin introduced into the medium of tissue cultures. J. Amer. Med. Assoc. 82:1609.

Goos, H.V. 1978. Hypophysiotropic centers in the brain of amphibians and fish. Amer. Zool. 18:401-410.

Grzedzielski, J. 1935. Zur Histologie der Rontgenkatarakt. Klin. Monat. Augenheilk. 95:360-369.

Hall, T.R. and A. Chadwick. 1979. Hyopthalamic control of prolactin and growth hormone secretion in different vertebrate species. Gen. Comp. Endo. 37:333-342.

Hammar, H. 1965. An autoradiographic study on cell migration in the eye lens epithelium from normal and alloxan diabetic rats. Acta Ophthal. 43:442-453.

Hanna, C. and J.E. O'Brien. 1963. Lens epithelial cell proliferation and migration in radiation cataracts. Rad. Res. 19:1-11.

Harding, C.V., A. Donn and B.D. Srinivasan. 1959. Incorporation of thymidine by injured lens epithelium. Exptl. Cell Res. 18:582-585.

Harding, C.V., W.L. Hughes, V.P. Bond and P. Schork. 1960. Audioradiographic localization of tritiated thymidine in whole-mount preparations of lens epithelium. Arch. Ophthal. 63:58-65.

Harding, C.V., J.R. Reddan, N.J. Unakar and M. Bagchi. 1971. The control of cell division in the ocular lens. International Review of Cytology 31:215-300.

Hayden, J.H. 1980. Doctoral Dissertation, University of Vermont, Burlington, Vermont.

Hayden, J.H. and J. Rothstein. 1979. Complete elimination of mitosis and DNA synthesis in the lens of the hypophysectomized frog: Effects on cell migration and fiber growth. Differentiation 15:153-160.

Hayden, J.H., H. Rothstein, B.V. Worgul and G.R. Merriam, Jr. 1980. Hypophysectomy exerts a radioprotective effect on frog lens. Experimentia 36:116-118.

Henle, J. 1882. Zur Entwicklungsgeschichte der Krystallinse und zur Theilung des Zellkerns. Archiv. F. Mikrosk. Anatomie, Bd. 20:413-430.

Hogben, L. 1923. A method of hypophysectomy in adult frogs and toads. Quart. J. Exptl. Physiol. 13:170-177.

Holder, A.T. and M. Wallis. 1977. Action of growth hormone, prolactin, and thyroxine on serum somatomedin-like activity and growth in hypopituitary Dwarf mice, J. Endocrinol. 74:223-229.

Howard, A. 1952. Whole mounts of rabbit lens epithelium for cytological study. Stain Technology 27:313-315.

Kibrick, E.A., H. Becks, W. Marx and H.M. Evans. 1941. Effect of different dose levels of growth hormone on the tibia of young

140

hypophysectomized female rats. Growth 5:437-447.

King, J.A., and R.P. Miller. 1979. Phylogenetic and anatomical distribution of somatostatin in vertebrates. Endocrinology 105:1322-1329.

Kistler, A., K. Yoshizato and E. Frieden. 1975. Binding of thyroxine and triiodothyronine by nuclei of isolated tadpole liver cells. Endocrinology 97:1035-1042.

Knapp, P. 1900a. Ueber Heilung von Linsenwunden beim Frosch. Z. Augenheilk. 3:209-228.

Knapp, P. 1900b. Ueber Heilung von Linserverletzungen beim Fisch. Z. Augenheilk. 3:50-56.

Kuhn, E.R. and H. Engelen. 1976. Seasonal variation in prolactin and TSH releasing activity in the hypothalamus of Rana temporaria. Gen. Comp. Endo. 28:277-282.

Leber, Th. 1878. Zur Pathologie der Linse. Zehender's Klin. Monatsbl., Beilageheft, Verh. der Heidelb. Ophth. Gesellsch. 16:33-47.

Mackenzie, D.S., P. Licht and H. Papkoff. 1978. Thyrotropin from amphibian. Rana catesbeiana. pituitaries and evidence for heterothyrotropic activity of bullfrog luteinizing hormones in reptiles. Gen. Comp. Endo. 36:566:574.

Martial, J.A., P.H. Seeburg, D. Guenzi, H.M. Goodman H.M. and J.D. Badter. 1977. Regulation of growth hormone gene expression: Synergistic effects of thyroid and glucocorticoid hormones. Proc. Natl. Acad. Sci. 74:4293-4295.

Mikulicich, A. and R. Young. 1963. Cell proliferation and displacement in the lens epithelium of young rats injected with tritiated thymidine. Invest. Ophthalmol. 2:344-354.

Messier, B. and C.P. Leblond. 1960. Cell proliferation and migration as revealed by radioautography after injection of ^{3}H-thymidine into male rats and mice. Amer. J. Anat. 106:247-285.

Phillips, L.S. and R. Vassilopoulou-Sellin. 1980. Somatomedins. New England Journal of Medicine 302:371-380; 438-446.

Poppe, E. 1924a. Experimental investigations of the effects of Roentgen-rays on the eye. Skr. Norske Vidensk.-Adak. Oslo, Mat. - Naturv. Kl:1-102.

Poppe, E. 1924b. Experimental investigations of the effects of Roentgen-rays on the epithelium of the crystalline lens. Acta Radiol. Stock. 23:354-367.

Rabl, C. 1900. Über den Bau und die Entwicklung der Linse. Leipzig Verlag von Wilhelm Engelmann.

Rafferty, N. 1963. Studies of an injury-induced growth in the frog lens. Anat. Rec. 146:299-312.

Rafferty, N. 1965. Propagation and prolongation of mitotic activity in the formation of injury-induced lentomas in Rana pipiens. Anat. Rec. 153:111-128.

Reddan, J.R., C.V. Harding, D. Harding, A. Weinsieder, N. Unakar, R. Shapiro and C. Mathews. 1975. Seasonal mitotic activity and wound healing in a teleost (Opsanus tau) ocular lens. Experienta 31:1026-1027.

Reddan, J.R., C.V. Harding, H. Rothstein, M.W. Crotty, P. Lee and N. Freeman. 1972. Stimulation of mitosis in the vertebrate lens in the presence of insulin. Ophthalmic Research 3:65-82.

Reddan, J.R., and D. Wilson. 1979. Insulin growth factors I and II stimulate mitosis in the epithelia of mammalian lenses cultured in a serum free medium. J. Cell Biol. 83:113a.

Rinderknecht, E. and R.E. Humbel. 1978. The amino acid sequence of human insulin-like growth factor I and its structural homology with proinsulin. J. Biol. Chem. 253:2769-2776.

Rinderknecht, E. and R.E. Humbel. 1978. Primary structure of human insulin-like growth factor II. FEBS Lett. 89:283-286.

Rosenbaum, D.M. and H. Rothstein. 1972. Mitotic variations in the lens epithelium of the frog. Ophthalmic Research 3:95-107.

Rothstein, H. 1968. Experimental techniques for investigation of the amphibian lens epithelium. In: Methods in Cell Physiology, ed. D.M. Prescott, pp. 45-74, vol. 3, Academic Press, NY.

Rothstein, H., E. Essner, K. Frank and A. Weinsieder. in press a. The metabolic basis for the lens growth inhibiting effect of hypophysectomy. II. The liver. Metabolic and Pediatric Ophthamology.

Rothstein, H. and C.V. Harding. 1962. Injury-induced synthesis of deoxyribonucleic acid in the lens of the sea bass. Nature 194:294-295.

Rothstein, H., R. Van Buskirk and J.R. Reddan. 1976. Hypophysectomy inhibits wound hyperplasia in the adult frog lens. Ophthalmic Research 8:43-54.

Rothstein, H., R. Van Buskirk, S.R. Gordon and Worgul, B.V. 1975. Seasonal variations in mitosis in the frog: a field study. Experienta 31:939-940.

Rothstein, H., J.J Van Wyk, J.H. Hayden, S.R. Gordon and A. Weinsieder. 1980. Somatomedin-c: Restoration in vivo of cycle traverse in GO/G1 blocked cells of hypophysectomized animals. Science 208:410-412.

Rothstein, H., N. Wainwright, S.R. Gordon and J.H. Hayden. 1979. Immunostaining of the bullfrog. Rana catesbeiana. lactotroph with antiserum to homologous prolactin. Cellular and Molecular Biology 26:1-7.

Rothstein, H., A. Weinsieder and R. Blaiklock. 1964. Response to injury in the lens epithelium of the bullfrog, Rana catesbeiana. Exptl. Cell Res. 35:548-556.

Rothstein, H., A. Weinsieder and M.E. Wilson. in press b. The metabolic basis for the lens growth inhibiting effect of hypophysectomy. I. The blood. Metabolic and Pediatric Ophthamology.

Sakharova, N.Yu. and V.A. Golichenkov. 1968. Seasonal changes in regeneration ability of frog. Rana temporaria. lens epithelium. Tsitologiya 10:896-899.

Salmon, W.D. and W.H. Daughaday. 1957. A hormonally controlled serum factor which stimulates sulfate incorpation by cartilage in vitro. J. Lab Clin. Med. 49:825-836.

Schalch, D.S., V.E. Heinrich, B. Draznin, C.J. Johnson and L.L. Miller. 1979. Role of the liver in mediating somatomedin activity: Hormonal effects on the synthesis and release of insulin-like growth factor and its carrier protein by the isolated perfused rat liver. Endocrinology 104:1143-1151.

Shapiro, B. and B. L. Pimstone. 1977. A phylogenetic study of sulphation factor activity in 26 species. J. Endocr. 74:129-135.

Sherrington, C.S. 1951. Man on his Nature, second edition, Cambridge University Press, London, New York City.

Snyder, B.W. and B.E. Frye. 1972. Physiological responses of larval and post-metamorphic Rana pipiens to growth hormone and prolactin. J. Exp. Zool. 179:299-314.

Streeten, B.W. and Eshaghian, J. 1978. Human posterior subcapsular cataract. Arch. Ophthal. 96:1653-1658.

Van Buskirk, R., B.V. Worgul, H. Rothstein and N. Wainwright. 1975. Mitotic variations in the lens epithelium of the frog. III. Somatotropin. Gen Comp. Endo. 29:52-59.

Vandesande, F. and K. Dierickx. 1980. Immunocytochemical localization of somatostatin-containing neurons in the brain of Rana temporaria. Cell Tissue Res. 205:43-53.

Van Wyk, J.J. and R.L. Hintz. 1979. Peptide growth factors. In: Endocrinology, ed. L.J. DeGroot, Grune and Stratton, New York, 1767-1775.

Van Wyk, J.J. and L.E. Underwood. 1975. Relation between growth hormone and somatomedin. Ann. Rev. Med. 26:427-441.

Van Wyk, J.J. and L.E. Underwood. 1978. The somatomedins and their actions. In: Biochemical Actions of Hormones. ed. G. Litwack, 5:101-148, Academic Press, New York.

Van Wyk, J.J., L.E. Underwood, M.E. Svoboda, D. Clemmons, D. Klapper, R.E. Fellows and H. Rothstein. 1980. Somatomedin-C: Chemistry and Biology. In: Growth Hormone and other Biologically Active Peptides, eds. A. Pecile and E.E. Muller. Excerpta Medica International Congress Series 495, Amsterdam, pp. 73-80.

Vassilopoulou-Sellin, R., L.S. Phillips and L.A. Reichard. 1980. Nutrition and somatomedin. VII. Regulation of somatomedin activity by the perfused rat liver. Endocrinology 106:260-267.

Von Becker, F.J. 1863. Untersuchungen uber den Bau der Linse bei dem Menschen und den Wirbelthieren. Archiv fur Ophthalmologie. 9:1-41.

Von Sallmann, L. 1957. The lens epithelium in the pathogenesis of X-ray cataract. 13th Edward Jackson Memorial Lecture. Am. J. Ophthal. 44:159-170.

Von Sallmann, L., P. Grimes and N. McElvain. 1962. Aspects of mitotic activity in relation to cell proliferation in the lens epithelium. Exptl. Eye Res. 1:449-456.

Von Sallmann, L. and P. Grimes. 1966. Effect of age on cell division, ^{3}H-thymidine incorporation and diurnal rhythm in the lens epithelium of rats. Invest. Ophthal. 5:560-567.

Wainwright, N., J.H. Hayden and H. Rothstein. 1978. Total disappearance of cell proliferation in the lens of a hypophysectomized animal: In vivo and in vitro maintenance of inhibition with reversal by pituitary factors. Cytobios 23:79-92.

Wainwright, N., H. Rothstein and S. Gordon. 1976. Mitotic variations in the lens epithelium of the frog. IV. Studies with isolated anuran pituitary factors. Growth 40:317-328.

Wedl, C. 1860. Atlas der Pathologischen Histologie des Auges Leipzig, Georg Wigan Verlag.

Weinsieder, A. 1973. Doctoral Dissertation, University of Vermont, Burlington, Vt.

Weinsieder, A., H. Rothstein and D. Drebert. 1973. Lenticular wound healing: Evidence for genomic activation. Cytobiologie 7:406-417.

Weinsieder, A. and H. Rothstein. 1980. Autoradiographic localization of human growth hormone in amphibian hepatocytes IRCS Medical Science 8:155-156.

Weinstein, R., B. Stemerman and T. Maciag. 1981. Hormonal requirements for growth of arterial smooth muscle cells in vitro: An endocrine approach to atherosclerosis. Science 212:818-820.

Worgul, B.V. and H. Rothstein. 1974. On the mechanism of thyroid mediated mitogenesis in adult anura: I. Preliminary analyses of growth kinetics and macromolecular synthesis, in lens epithelium, under the influence of exogenous triiodothyronin Cell and Tissue Kinetics 7:415-424.

Worgul, B.V., G.R. Merriam, Jr., A. Szechter and B.D. Srinivasan 1976. The lens epithelium and radiation cataract. I. Preliminary studies. Arch. Ophthal. 94:996-999.

143

Worgul, B.V., H. Rothstein, C. Medvedovsky, G.R. Merriam, Jr. and
    M.E. Wilson. in press. Radiation cataractogenesis in the
    amphibian lens. Ophthalmic Research.
Yalow, R.S., K. Hall and R. Luft. 1975. Radioimmunoassay of
    somatomedin B: Application to clinical and physiologic
    studies. J. Clin. Invest. 55:127-137.
Zapf, J., E. Rinderknecht, R.E. Humbel and E.R. Froesch. 1978.
    Nonsuppressible insulin-like activity. NSILA. from human
    serum: Recent accomplishments and their physiologic
    implications. Metabolism 27:1803-1828.

Coordinate Control of Retinal Neovascularization

Robert J. Mello

Angiogenesis refers to the development of new blood vessels. Such a development can occur as a response to either normal physiological events or abnormal pathological conditions. Examples of normal angiogenic processes include the vascular development which occurs during embryogenesis (Sabin, 1920; Wagner, 1980) and wound healing (Clark and Clark, 1932; Cliff, 1963; Schoefl, 1963). The precisely controlled vascularization which occurs during embryogenesis is crucial to proper fetal development, not only in providing for nutrient and waste exchange within rapidly developing tissues, but also, perhaps, in influencing the development of surrounding tissues (Goudie et al., 1980). Using histological techniques, embryologists have documented the transformation of embryonic mesodermal cells into blood islands and angioblasts (Sabin, 1920; Wagner, 1980). The ability to form new vessels is not lost with the cessation of embryogenesis. Indeed, if influenced by an appropriate stimulus, an angiogenic response can be induced in most post-embryonic vascularized tissues. During the process of wound healing, for example, damaged tissue, including vascular tissue, is either repaired or replaced (Cliff, 1963; Schoefl, 1963). However, unlike the mesodermal-derived vascularization events of early embryogenesis, most neovascularization events of fully developed tissues occur as extensions of nearby, pre-existing blood vessel networks.

As mentioned, neovascularization also occurs as a prominent feature in a variety of pathological conditions such as solid tumor growth (Folkman and Cotran, 1976), psoriasis (Folkman, 1972), inflammation (Schoefl, 1963), and a variety of vascular proliferative retinopathies (Henkind, 1978). Again, much of the

past body of information concerning new vessel growth in these
cases has been obtained through the combination of macroscopic
observation and microscopic, histological technique. In all of
these pathologies, the data have revealed, in essence, that new
blood vessels have formed subsequent to cell or tissue damage,
tumor growth, inflammation, or ischemia. Presently, angiogenesis
research is in its most exciting and potentially beneficial stage.
How does neovascularization occur? What are the cellular and/or
soluble factors (angiogenesis factors) which control the induction
of a neovascular response from the adjacent vasculature? Can
pathological angiogenesis be interrupted or inhibited by
antiangiogenic agents? The morphological studies of Yamagami
(1970) and Ausprunk and Folkman (1977) have detailed the cellular
events involved in the formation of new capillary sprouts from
pre-existing vessels. The existence of soluble, diffusible
angiogenic factors was demonstrated for tumor-induced angiogenesis
by Greenblatt and Shubik (1968) and for lymphocyte-induced
angiogenesis by Muthukkaruppan and Auerbach (1979). An excellent
review of the many factors which have been associated with
angiogeneic responses (especially lymphocyte-induced angiogenesis)
was recently presented by Auerbach (1981). In addition, intense
study of tumor-induced neovascularization has implicated a low
molecular weight substance from the Walker 256 rat carcinoma in
endothelial cell mitogenesis (McAuslan and Hoffman, 1979; Weiss et
al., 1979; Fenselau et al., 1981). However, only the purified,
low molecular weight substance of Fenselau et al. (1981) has been
shown to be not only mitogenic for endothelial cell in vitro but
also angiogenic in vivo in the rat corneal micropocket bioassay
(Fournier et al., 1981). In a comprehensive review of the
relationship between tumor growth and tumor vascularizaton,
Folkman and Cotran (1976) not only summarize tumor-induced
angiogenesis but also explain the concept of anti-angiogenesis.
Studies of the control of tumor growth (or other pathological
angiogenic events) through the inhibition or reversal of the
neovascular response would seem to be the most clinically
beneficial aspect of current angiogenesis research.

Neovascularization also occurs in a variety of ocular
pathologies which can involve the cornea, iris, retina, or
choroid. Reviews of ocular neovascularization have been presented
by Wise (1956) and more recently by Henkind (1978). It should be
mentioned at this point that the factors which control ocular

angiogenesis may well be different from those involved in tumor angiogenesis. This paper will describe some of the current efforts which have been performed to understand the factors involved in the regulation of retinal neovascularization.

To study the biochemical factors involved in the control of retinal blood vessel proliferation, assay systems capable of producing and monitoring neovascularization are required. Since suitable animal models of diabetic retinopathy do not exist, it is not yet possible to study this form of retinal vascular proliferation. The puppy model of oxygen-induced experimental retrolental fibroplasia, developed by Patz et al (1953), is capable of producing retinal neovascularization. However, it is wholly unsuitable for use as a routine assay in a standard biochemical purification protocol due to the excessive cost, low sample capacity, and long time factors which would be involved. Ideally, the most efficient angiogenesis assay would be one which possessed high sample capacity, speed, sensitivity, low cost, and a readily observable neovascular response. Unfortunately, this ideal system does not exist. However, after studying the systems used to investigate tumor angiogenic factors, it is possible for a combination of assays to possess all of the qualities desired in the "ideal" angiogenesis assay. Listed in Table I are the assays which are now routinely used in the purification of a retina-derived angiogenic factor and a vitreous-derived anti-angiogenic factor. Both the rabbit corneal micropocket (Gimbone et al, 1974) and the chicken chorioallantoic membrane, or CAM (Ausprunk et al, 1975) are in vivo assays which are used to demonstrate the stimulation or inhibition of new blood vessel growth. Cell cultures (Fenselau and Mello, 1976; Mello, 1977; Glaser, 1980b) are used to monitor the in vitro effects of various test fractions on the proliferation or migration of vascular endothelial cells. The in vitro assays are more rapid, and can handle a greater number of samples at a lower overall cost.

Table I

Basic Assays Used in Angiogenesis Studies

1. Rabbit corneal micropocket implant
2. Chicken chorioallantoic membrane implant (CAM)
3. Cultured vascular endothelial cells
    A) Proliferation assay
    B) Migration assay

However, the focus of the research is on angiogenesis, and for that reason the in vivo bioassays are critical.

The rabbit corneal micropocket technique was developed by Gimbrone et al., (1974) for the study of tumor-induced neovascularization. Potential stimulators of vessel growth, of either tumor or retinal origin, can be implanted in the cornea approximately 1-2mm from the vasculature of the limbus. The stimulus diffuses through the cornea and new vessel growth, directed toward the stimulus, can be observed and quantitated. The basic technique is as follows: An incision is made on the cornea of an anesthetized rabbit eye. A small cyclodialysis spatula is inserted into the incision and advanced to within 1-2mm of the limbus, thus forming a pocket in the cornea. Test materials are placed within and advanced to the front of the pocket. The incision is then sealed by applying mild pressure with the side of the spatula. An early positive response can be observed 5 to 7 days after implantation, and by slit-lamp microscopy usually appears as an edge of neovascular tufts advancing toward the implant. This system is quantified by measuring the mean vessel length of the advancing capillary front as a function of time.

The other angiogenesis assay which we routinely perform utilizes the rich vascular bed of the chicken chorioallantoic membrane (CAM). Angiogenic materials implanted on the CAM will elicit new vessel growth directed toward the implant. Six-day-old fertilized chicken eggs first are candled to locate the air sac and the embryo. A window is carefully cut in the shell and the displacement of the air sac allows the CAM to fall away from the shell membrane. This membrane is then removed, the window sealed with tape, and the egg is returned to the incubator. Two days later the angiogenic stimulus which has been encased in a slow release Elvax pellet (Langer and Folkman, 1976) is placed on the CAM, and a sterile plastic coverslip is placed over it. This coverslip serves a two-fold purpose. First, it helps prevent the pellet from being moved by the embryo. Second, it serves as the vehicle for the delivery of inhibitory substances in the anti-angiogenesis assays. The window is sealed again and the egg is returned to the incubator. New vessel growth is monitored visually and photographically 4 to 7 days after implantation of the stimulus (Lutty, et al., 1981b). Assessment of vessel growth is made qualitatively on a 0-4+ scale which is a modification of that described by Folkman and Cotran (1976) and which ranges from a 0 response showing no apparent organization of the vessels with respect to pellet, to a 4+ response in which numerous arcs, fringes of vessels, and patent vascular channels are observed on the pellet. An example of 4+ stimulation is shown in Figure 1. The numerous vessel arcs and vascular channels lying over the implant can be readily seen in this India ink preparation.

On a routine basis, we prefer to use the CAM assay for the preliminary screening of potential angiogenic or antiangiogenic substances. The CAM assay produces results in less time; that is, an average of less than one week, compared to as much as one month with the corneal assay. There is also less risk of interference from the host immune system with the CAM. Finally, the CAM assay is less expensive to perform than the corneal assay. However, even the CAM assay has its limitations. It is a difficult system to define and control because it is, by nature, a rapidly developing embryonic system. Also the method of grading is

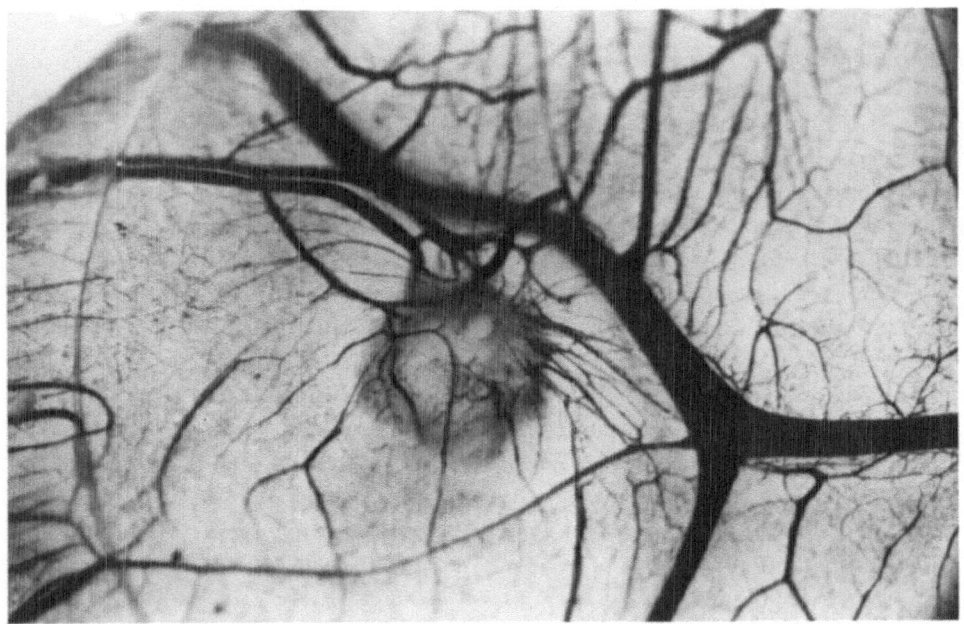

Figure 1. Demonstration of 4+ neovascularization on the chicken CAM. India ink preparation taken seven days after implantation of an Elvax pellet containing retina-derived angiogenic factor. Note numerous small vessel loops situated over the pellet.

qualitative rather than quantitative.

Clearly, even simpler, more rapid assay systems would be required to facilitate the purification and characterization of the factors involved in the control of new vessel formation. We, therefore, turned to the in vitro techniques of cell culture. Since we were dealing with developing vasculature, the cell of choice was the vascular endothelial cell.

As a result of the work of Yamagami, and Ausprunk and Folkman, it is known that neovascularization results from a process of sprouting from pre-existing vessels (Yamagami, 1970; Ausprunk and Folkman, 1977). This process is thought to occur in three stages. First, there is an apparent rupture of the basement membrane. This would require some type of collagenolytic activity. This break in the basement membrane would then allow endothelial cell migration to occur from the tip of the new vessel toward the neovascular stimulus. This would most likely involve chemotactic factors. Finally, endothelial cell replication must occur for the extension of the vessel wall and this would most likely require the involvement of mitogenic factors.

Utilizing cultured endothelial cells, systems have been developed which are capable of assaying both migration (Glaser, et al., 1980b) and proliferation phenomena (Fenselau and Mello, 1976; Mello, 1977). Cells are isolated from fetal bovine aortas by everting the aorta, thereby exposing the endothelium. The fetal bovine aortic endothelial (FBAE) cells are then removed by treatment with proteolytic enzymes under controlled conditions. The stimulation or inhibition of cellular proliferation in response to added test substances is measured in sparse cultures of FBAE cells either by ^{3}H-thymidine incorporation or by direct counting of the cells. The proliferative (or antiproliferative) capacity of the sample is then compared to buffer controls. As seen in Figure 2, a linear relationship between ^{3}H-thymidine incorporation and cell density does exist over a range of cell densities. Within this range either method of quantitation may be used, and under our controlled assay conditions, there is good agreement between the two methods (Mello, 1977).

Migration and chemotaxis are measured in a modified Boyden chamber consisting of a bottom well containing the chemoattractant, and an upper chamber into which cells are dispensed. In between the two chambers is a polycarbonate membrane containing 5 μ pores through which the cells can migrate.

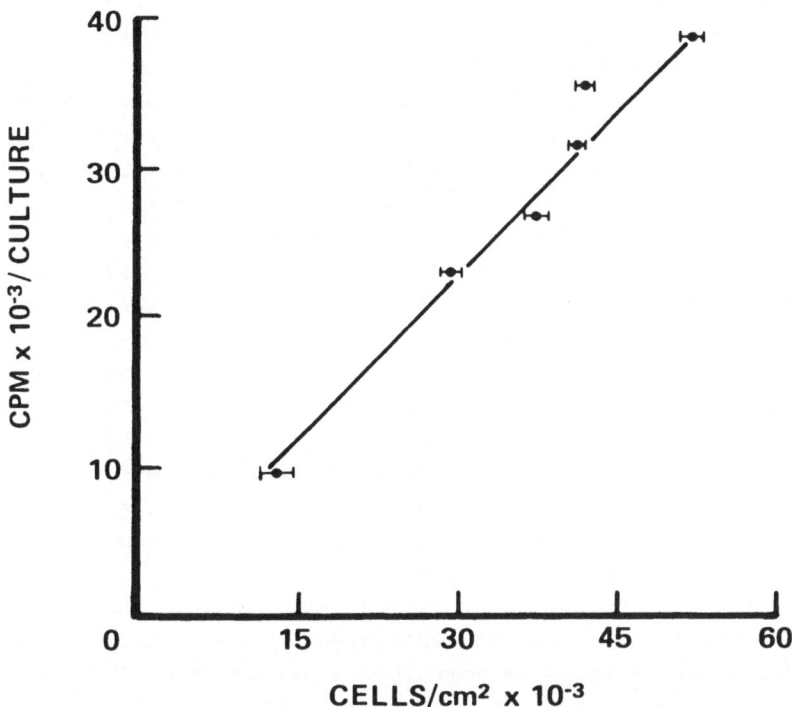

Figure 2. Incorporation of ^{3}H-thymidine (^{3}H-Tdr) as a function of cell density in actively growing cultures of fetal bovine aortic endothelial (FBAE) cells. Passage 8 cells were plated at 15x10[3] cells/cm[2] (Linbro 24 well plates) and were allowed 3 hours to attach. Growth media were then changed to 1.5 ml/well of Medium 199 containing O% - 10% fetal bovine serum (FBS) as growth stimulus. After 48 hours, one set of cultures received a 1 hour pulse of ^{3}H-methyl-Tdr (O.625 mCi, 6.7 Ci/mmole). Incorporation of radiolabel into acid insoluble material was determined following acetic acid-ethanol fixation and perchloric acid extraction of the cells. Cell density was determined in a replicate set of cultures by resuspending the cells with trypsin/EDTA solution and enumerating them electronically in a Coulter counter. Results indicated that, under these conditions, increases in ^{3}H-Tdr incorporation were linearly related to cell density up to 50-60 x 10[3] cells/cm[2].

After a standardized time period the cells above the membrane are physically removed with a cotton swab, and the cells attached to the lower face of the membrane are counted microscopically following fixation and staining (Glaser, et al., 1980b).

Having detailed these assay systems, I would now like to show you how they have been used in studies of a retina-derived angiogenesis factor which produces stimulatory effects in these systems and a vitreous-derived antiangiogenesis factor which produces inhibitory effects. The following is a brief summary of the work which has been done on the retinal angiogenic factor (RAF) by D'Amore, et al., (1981) and Glaser, et al. (1980a,b).

The starting material for the purification of the RAF is a balanced salt extract of adult bovine retina. Angiogenic activities have been demonstrated in extracts of bovine, feline, and human retinas (Glaser, et al., 1980a). In each case the crude salt extracts were initially screened in the cell culture assays. Results of the cell culture assays revealed that the retinal extracts were mitogenic for vascular endothelial cells and corneal fibroblasts. In each case the growth stimulatory response was dose dependent. The ability of the retinal extract to stimulate the in vitro growth of corneal fibroblasts suggests that a retinal-derived factor may be implicated in the proliferation of the fibrous tissue which is sometimes observed in conjunction with intraocular neovascularization (Yanoff, 1969). Retinal extract does not appear to be a general mitogen for all cell types since no cell proliferative effects were observed when vascular smooth muscle cells were employed in the assay.

In another study (Glaser et al., 1980b) utilizing the migration assay, bovine retinal extract was shown to possess true chemotactic activity for fetal bovine aortic endothelial cells in culture. That is, these cells respond to a concentration gradient of the retinal factor by migrating in the direction of an increasing concentration of the factor.

Angiogenic activity of the retinal extract was determined by its ability to induce new blood vessel formation on the chicken chorioallantoic membrane and also in the rabbit corneal micropocket assay. Based upon the number of new capillary vessel loops which were formed and directed toward the implanted retinal extract, it was concluded that retina contains a substance(s) which, in addition to its in vitro cell proliferative and chemotactic activities, can markedly stimulate new blood vessel formation in vivo.

D'Amore et al., (1981) have effected a partial purification of the active factor. Bovine retinas were used as the starting material since sufficiently large quantities of relatively fresh (2-3 hours post-sacrifice) retinas could be obtained. Briefly, retinas were extracted in a phosphate buffered balanced salt solution for three hours at room temperature. Using dilute acid, the extract was then brought slowly to pH 4. After removal of the inactive precipitate by centrifugation, the supernatant was neutralized, dialyzed against 50 mM Tris-HCl buffer (pH 7.2) and applied to an ion-exchange column of DEAE Bio-Gel A. The active fraction was bound to the column under these conditions, and could be eluted by the addition of 100 mM sodium chloride to the buffer.

Biochemically, D'Amore et al., (1981) offer the following characterization of the retinal angiogenic factor. The activity appears to be associated with an anionic molecule in that it was bound to a positively charged ion-exchange matrix at neutral pH. The active fraction was non-dialyzable and based upon the results of gel filtration (Sephadex G-50) and ultrafiltration experiments, the molecular weight is thought to be between 50,000 - 100,000 daltons. Sodium dodecyl sulfate (SDS) polyacrylamide slab gel electrophoresis of the most purified active fraction revealed the presence of two major components (MW_r = 50,000 and 70,000), as well as a variable amount of minor components of lesser size. The active factor was stable to treatment with reducing agents, DNAase, RNAase, and mild heating (56°C for 30 minutes). It was also stable to changes in hydrogen ion concentration between pH 4.0 - pH 9.0. However, extreme shifts of pH (below pH 2 or above pH 12) resulted in loss of activity. It was also reported that activity was lost following treatment with 0.02% (W/V) SDS, heat (at temperatures greater than 65°C), and prolonged pronase treatment.

It was inferred from these, and other characteristics (D'Amore et al., 1981) that the active material(s) in the retinal extract is a protein macromolecule. This differs from the reports of others (Fenselau et al., 1981; McAuslan and Hoffman, 1979; Weiss et al., 1979), which suggest much smaller molecular weights for a variety of other angiogenic substances. This macromolecular nature of the retinal angiogenic factor may allow it to exert its effect in a manner which could be unique to retinal tissue. If so, then this uniqueness could be used to advantage in the

treatment of retinal neovascularization. Perhaps, by introducing
an antiangiogenic factor which could interfere with the activity
of the retinal angiogenic factor, pathological retinal
neovascularization could be controlled or prevented. One possible
source of an ocular antiangiogenesis factor(s) is the normal adult
vitreous of the eye.

The initial observation which suggested that an angiogenesis
inhibitor might be present in vitreous stemmed from observations
made on the growth of intravitreal tumor implants in rabbits (Brem
et al., 1976). Whereas tumors implanted in the cornea or on the
iris would readily induce new vessel growth at a distance of
1-2mm, it was found that intravitreal tumors remained avascular
and that the induction of no new vessels occurred. This apparent
inhibition of neovascularization occurred even if the tumors were
located as little as O.1mm from the retina. Only when the tumor
was in direct contact with the retina did it become vascularized
and grow rapidly. From this it was postulated that a substance
may be present in the vitreous which inhibited tumor-induced
neovascularization, and perhaps neovascularization in general.
Preliminary tests with isolated rabbit vitreous assayed in the
rabbit corneal micropocket further documented the hypothesis (Brem
et al., 1977). We are now in the process of purifying the
inhibitory factor from a more available bovine vitreous source.

Cow eyes were obtained from a local abbatoir within three
hours of slaughter. The vitreous was carefully removed and
homogenized. The viscous sample was then centrifuged at high
speed, dialyzed extensively against water and lyophilized. The
final lyophilized material can either be incorporated into Elvax
(ethylene-vinyl acetate copolymer) slow-release pellets for
corneal implantation, or redissolved in various aqueous solutions
for testing in our other assay systems.

When this crude vitreal preparation was tested in the in
vitro endothelial cell migration assay (Table II) (Lutty et al.,
1981a), a dose-dependent inhibition of retinal angiogenic
factor-induced cell migration was observed. Not only was cell
migration affected, but vitreous extracts also inhibited the
proliferation of fetal bovine aortic endothelial cells in culture.
In Figure 3, fetal bovine serum is used as the growth stimulus.
It can be seen that increasing doses of vitreous inhibited cell
proliferation, here monitored either by cell counting or by the
incorporation of ^{3}H-thymidine into nuclear DNA. The cells used in

Table II

Inhibition of Cell Migration by Bovine Vitreous Extract

Addition[a]	Cell migration[b] (mean ± S.D.)	Net cell migration[c]	% Inhibition of net RAF[d]-induced migration
Control (no addition)	42 ± 10	- 0 -	-
Bovine RAF (32 µg/ml)	192 ± 34	150	-
Bovine RAF (32 µg/ml) + Bovine vitreous			
100 µg/ml	103 ± 13	61	59
300 µg/ml	83 ± 12	41	73
500 µg/ml	64 ± 19	22	85

[a]Additions were made to Minimum Essential Medium containing 10% fetal bovine serum.

[b]The number of cells observed in 20 oil immersion microscopic fields.

[c]Test cell migration - Control cell migration.

[d]RAF: Retina-derived angiogenic factor.

these proliferation assays were previously labeled with low levels of ^{14}C-thymidine. If the decreases in cell number or ^{3}H-thymidine incorporation were due to toxic substances in the vitreous, then there would have been a significant loss of ^{14}C-labeled cells from the assay. No such losses occurred. Even at the highest dose levels where cellular proliferation was decreased 50-65%, there was less than an 8% loss in ^{14}C-labeled cells. Therefore, in the dose ranges tested to date, the bovine vitreous extracts appear to be non-toxic to cultured fetal bovine aortic endothelial cells.

Since the cow eyes remained intact for about three hours before the vitreous was removed, the possibility existed that the non-perfused retinas were producing retinal angiogenic factor (RAF) which could diffuse into the vitreous. Since RAF was heat labile, this "contaminant" of the vitreous could be inactivated by a simple heat treatment. Previous studies in our laboratory had indicated that the antiangiogenic activity of crude vitreous was heat stable. (It should be noted that the vitreous extracts tested

156

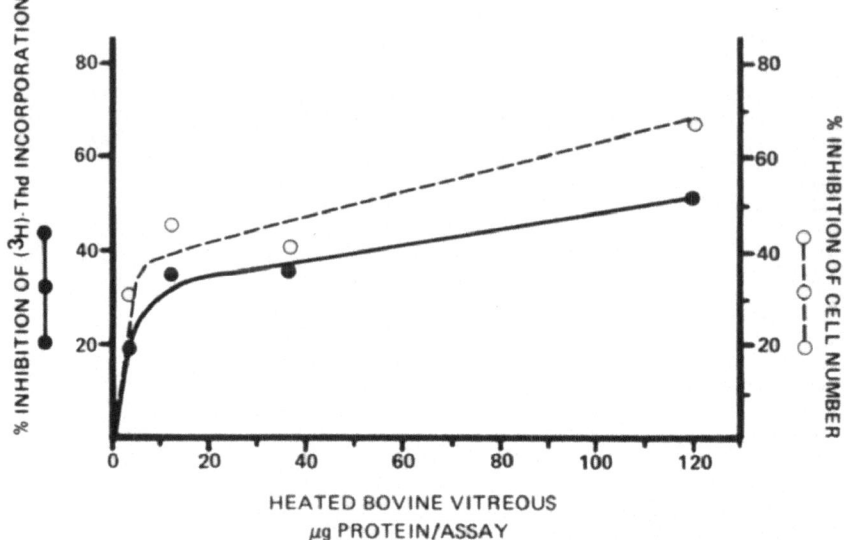

Figure 3. Inhibition of FBAE cell proliferation by bovine
vitreous. FBAE cells were plated at 10^4 cells/well and allowed 3
hours to attach before the medium was changed to 1.5 ml/well
Medium 199 containing 1.5% FBS and varying amounts of heat-treated
bovine vitreous protein. Cells were incubated for 44 hours at
37^OC. The extent of cell proliferation was determined in
replicate cultures using the ^{3}H-Tdr incorporation assay (●) or
the cell counting assay (○). Results are expressed as a
percent inhibition of cell proliferation compared to cultures
which received buffer additions which lacked vitreous.

in Figure 3 were heat-treated at 95^OC for 10 minutes.)

We therefore decided to assay both heated and non-heated
vitreous extract for their effects upon cellular proliferation
induced either by serum or RAF. The results of one such assay are
shown in Figure 4. In the presence of serum alone, non-heated
vitreous produced up to a 53% stimulation of ^{3}H-thymidine
incorporation. When a maximal stimulatory dose of exogenous RAF
was included in the assay medium, non-heated vitreous had no
apparent effect. However, heat-treating the vitreous converted it
into a potent inhibitor of both serum-induced and RAF-induced cell
proliferation.

It appeared, then, that there was a heat-labile growth
stimulatory factor present in these vitreous samples. Based upon

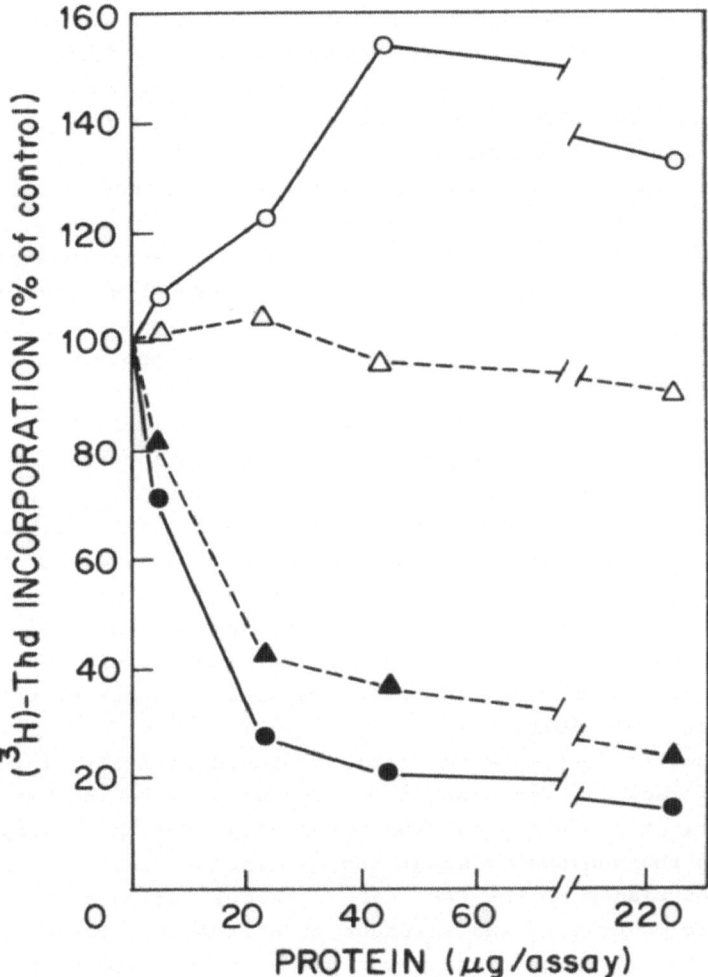

Figure 4. Effect of heat treatment of bovine vitreous on its
ability to inhibit the serum-induced or serum plus retinal
factor-induced proliferation of FBAE cells. Cells were plated as
in Figure 3. Lyophilized bovine vitreous was dissolved in 50 mM
Tris-HCl, pH 8.0, at 6 mg/ml (dry weight) and then 0.22 μ filter
sterilized. One aliquot was heated at 95°C for 10 minutes, then
cooled on ice prior to addition to the assay medium. Assay medium
was Medium 199 containing 1.5% FBS in the absence (————) or
presence (– – –) of retina-derived angiogenic factor (RAF, 125
g/ml, Lowry protein). This level of RAF produced a 2-fold
stimulation of 3H-Tdr incorporation when compared to control
cultures which lacked both RAF and vitreous proteins. Vitreous
samples, either non-heated (O   Δ) or heat treated (●   ▲) were
added in the amounts shown (Lowry protein). Cells were pulse
labeled for 1 hour with 3H-Tdr after 44 hours at 37°C.

its varied level of activity and its heat lability, it was assumed that this low level of stimulation was the result of contamination by RAF which had diffused into the vitreous during the time period between slaughter of the animal and the separation of vitreous from retina in the laboratory.

The inhibition of cell proliferation by vitreous was also observed to be reversible as seen in Figure 5. These are parallel cultures of fetal bovine aortic endothelial cells stimulated to grow by serum. Panel A shows a control culture three days after the start of the experiment. Panel B depicts a parallel culture grown for the same three-day period, but 100 μg protein/ml of heat-treated vitreous extract was added at the start of the test. It is obvious that the vitreous-treated cultures contain fewer cells. If the culture in Panel B is allowed to incubate an additional three days, the number of cells slowly begins to increase. The growth rate of these cultures is still markedly reduced. If a three-day vitreous-treated culture, such as in Panel B, is washed and returned to normal medium containing 10% serum for an additional period of three days (Panel D), the cells return to their normal growth rate and repopulate the culture dish. Therefore, it appears that the cell growth inhibitory effects of vitreous are reversible.

In order to determine the antiangiogenic activity of the vitreous extract, it was assayed for its ability to inhibit neovascularization in the chicken CAM assay. For this assay, RAF was used as the angiogenic agent and was encased in a slow-release Elvax pellet placed on the CAM. The vitreous, dried to the bottom of a sterile coverslip, was released by diffusion. The coverslip containing the vitreous extract was placed directly over the RAF-Elvax pellet. Results of such an assay are given in Table III (Lutty, et al., 1981b). The mean growth score refers to the qualitative 0-4[+] grading system mentioned earlier. The higher the score, the greater the neovascular response. It can be seen that in the absence of vitreal extract there was significant new vessel growth induced by RAF. However, in the presence of vitreal extract, this response was greatly diminished.

Vitreous extract has also been tested in the rabbit corneal micropocket assay. The rabbit cornea is very sensitive to the implantation of xenogenic tissue, and it reacts strongly by mounting an inflammatory response against the implant. For that reason a rabbit V-2 carcinoma was used as the angiogenic stimulus.

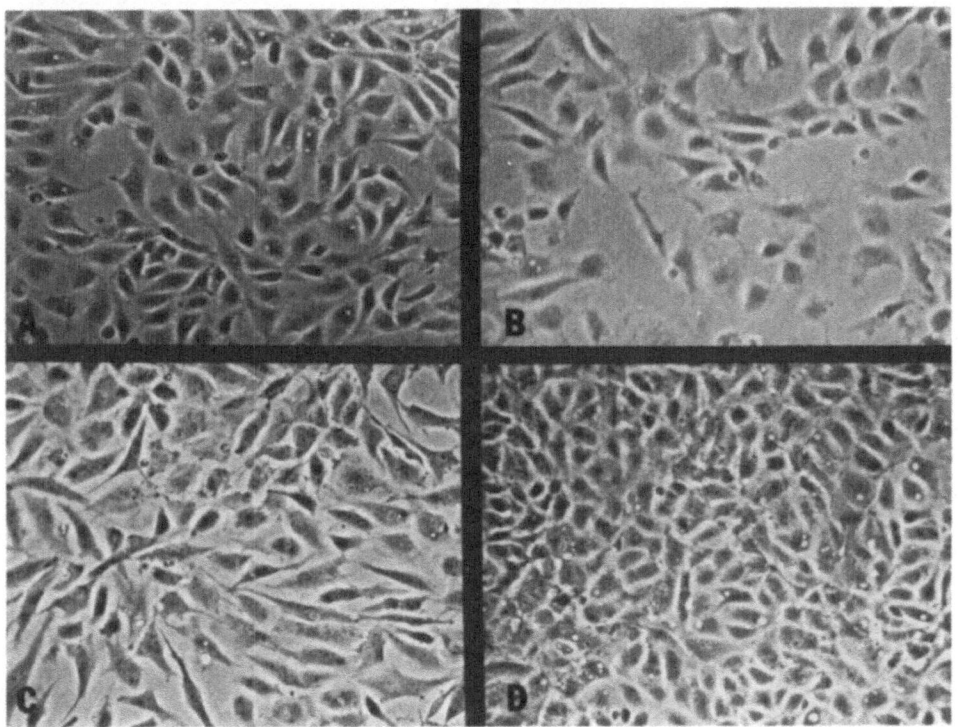

Figure 5. Reversible inhibition of FBAE cell growth produced by
heat-treated bovine vitreous extract. Subconfluent cultures of
FBAE cells in 6 cm dishes were incubated in 5 mls of Medium 199
containing 1% FBS. Panel A) Control culture (buffer addition
only) after 3 days. Panel B) Parallel culture 3 days after the
addition of 500 µg of heat-treated vitreous protein. Panel C)
Parallel culture 6 days after addition of 500 µg of vitreous
protein (no medium change). Panel D) Parallel culture after an
initial 3 day treatment with 500 µg of heat-treated vitreous,
followed by replacement of the medium with standard medium
containing 10% FBS in the absence of vitreous protein (total
incubation time, 6 days). Phase contrast x350.

Rabbit vitreous extract was encased in Elvax and interposed

between the tumor and the limbal vessels. Control Elvax pellets

lacking vitreous were implanted in the other rabbit eye. After 14

days there was extensive neovascularization in the control eye,

and new blood vessels readily formed over the blank Elvax pellet.

In the eye receiving Elvax containing vitreous extract,

Table III

Inhibition of Neovascularization of the Chicken CAM

Time[a]	Sample[b]	Mean Growth Score
1	Water	0.10
	200 μg vitreous	0.00
2	Water	1.89
	200 μg vitreous	0.79
3	Water	2.66
	200 μg vitreous	0.82

[a]Days after implantation of pellet and coverslip.

[b]Samples of either water or vitreous extract in water were filter sterilized (.22 μ filters) and 90 μl aliquots were applied to each coverslip and allowed to air dry under aseptic conditions.

neovascularization, while not completely inhibited, was markedly reduced. Also, the vessels which formed had a tendency to grow around, rather than over, the vitreous implant.

The purification of the vitreous inhibitor is in progress. Some of the characteristics of the crude extract are its stability to heat (95°C for 10 minutes), acid (to pH 1.5), and hyaluronidase treatment. It is non-dialyzable with an apparent molecular weight greater than 50,000 daltons. It inhibits retinal and tumor factor-induced angiogenesis in the chicken CAM and rabbit cornea assays. In vitro it inhibits the migration and proliferation of fetal bovine aortic endothelial cells in a reversible, non-toxic manner.

The presence of two substances (RAF and vitreous antiangiogenic factor) possessing such opposite effects in close proximity to each other suggests that they may be acting in a coordinate manner to regulate the normal vascular state of the retina. If these factors are acting in tandem, then fluctuation in their local concentrations, either by increases in retinal factor production or by decreases in vitreous inhibitor levels, might result in the development of abnormal retinal neovascularization.

DISCUSSION

By studying the pathology of the retinopathies, some insight into their cause(s) may be obtained. These insights can then be

channeled into the development of assay systems capable of
monitoring various cellular and/or molecular aspects of the
neovascular response. However, this approach may not always yield
immediate results. For example, in studying the pathological
effects of diabetic retinopathy - a major cause of blindness - one
observes that the "proliferative stage" of the disease is
characterized by marked retinal neovascularization. Yet, more
in-depth studies into the mechanisms responsible for the
development of diabetic retinal neovascularization have been
severely hampered by the lack of suitable animal models for the
disease. Until such model systems are developed, we must rely on
indirect clues derived from the study of other systems. One such
system is experimental retrolental fibroplasia (RLF) for which
good animal model systems do exist.

In humans, RLF is the retinopathy most associated with the
premature infant. Although animal models developed by Ashton et
al (1953) and Patz et al (1953) utilized newborn rather than
premature animals, they took advantage of the fact that the
immature retinal vasculature of the newborn kitten or puppy
closely resembles the immature retinal vasculature of a human
fetus of approximately seven months gestational age. The
pathogenesis of RLF proceeds through two stages. The primary
stage involves an oxygen-induced constriction of the immature
retinal vessels. This response to oxygen is a function of both
oxygen concentration and the duration of treatment (Patz, 1980).
Upon mild oxygen treatment, only the peripheral vessels are
affected and neovascularization subsequently occurs at that point.
With increased oxygen exposure, neovascularization occurs more
posteriorly, such that, with very high oxygen exposures, the
capillary bed near the disc becomes affected. The secondary stage
of the disease occurs upon return to room air. The
vasoconstriction is relieved, intact vessels recanulate, and new
vessel growth may not follow its original path and may break
through the plane of the retina into the vitreous compartment.
Since many of these new vessels may not be structurally intact,
hemorrhages can result. Blindness occurs when the scar tissue
that subsequently develops places traction on the retina,
eventually resulting in retinal detachment.

Among the first conclusions to be drawn from the experimental
animal models of RLF was that exposure of the new born kitten or
puppy to high levels of oxygen resulted in a constriction of the

retinal blood vessels. Vasoconstriction could produce decreased blood flow through the vessels which then may have resulted in local areas of retinal ischemia. Neovascularization following ischemia has since been observed in a number of other proliferative retinopathies including diabetic retinopathy, branch vein occlusion, and the sickle hemoglobinopathies. In most instances it appears that neovascularization occurs at the border between perfused and non-perfused areas of the retina. Recent observations of the proliferative retinopathies, combined with the earlier work on normal retinal vascularization and abnormal retinal neovascularization (Michaelson, 1948; Ashton et al., 1953), have led to the development of an hypothesis which suggests that areas of non-perfused, ischemic retina may elicit a diffusible "factor X" which may induce neovascularization either next to the non-perfused area, at the iris, or at the optic disc (Patz, 1980). Where the neovascularization occurs will depend, as Henkind has suggested, on the sensitivity of the different vascular beds to the stimulus, and also on the overall disease state of the retinal vessels (Henkind, 1978).

The exact nature of the vasoproliferative "factor X" is as yet unknown. It may represent a series of events such as the prostaglandin cascade which has been postulated to play a role in the pathogenesis of experimental RLF in puppies (Flower et al., 1981). Perhaps it is more likely to be a discrete entity such as one of the vitreous soluble proteins whose sustained presence in puppy eyes appeared to correlate with elevated oxygen administration (Chen and Patz, 1976). However, no indication of the source or nature of these vitreous soluble proteins was given. The angiogenic substance described by D'Amore et al (1981) is derived from the mature retina and is at least partially characterized. Whether this is the factor responsible for inducing the neovascularization which occurs in retinal vascular diseases is as yet unknown. Each of the previous examples represents an agent which could exert a direct, postitive effect on the retinal vasculature. Consideration must also be given to the possibility that these agents could act indirectly upon the vasculature by removing or inactivating an endogenous inhibitor of vessel proliferation. In essence, the vessels may normally be under the influence of anti-proliferative factors which prevent random neovascularization. What may develop, therefore, is a system of balance between positive and negative vascular effectors which would be capable of exerting a greater degree of control over vascular proliferation. The vitreous antiangiogenic factor described earlier may be such a negative vascular effector.

From the standpoint of embryonic development, it is interesting to note tht the primary embryonic vitreous is highly vascularized (Jack, 1972). The vasculature is composed of the hyaloid system and the tunica vasculosa lentis, the primary function of which is to provide a blood supply for the developing lens. The vascularized primary vitreous appears toward the end of the third week of human development and by the beginning of the 12th week of development it has already begun to be replaced by the more permanent secondary vitreous. Occurring simultaneously with the formation of the secondary vitreous is the degeneration of the hyaloid vasculature and the tunica vasculosa lentis. Such continued degeneration ultimately results in the formation of an essentially avascular vitreous (Duke-Elder, 1963).

The mechanisms involved in the degeneration of the hyaloid system are not fully understood. One possibility is that regression of this embryonic vasculture of the vitreous might result from the loss of factors necessary for the continued proliferation of endothelial cells, or for the maintenance of their integrity. The decreased production of angiogenesis factors or, perhaps, their neutralization by increasing levels of newly formed antiangiogenic substances in the secondary vitreous might also produce a regression of the embryonic vasculature.

Therefore, considering both the normal ocular vascularization of development and the abnormal neovascularization which occurs during proliferative retinopathies, the control of ocular angiogenesis may result not only from a direct stimulatory action on vascular tissue, but also from an indirect release from inhibition. Perhaps ischemic retina leads to neovascularization, as Wise suggested, due to the production once again of some fetal-life vasoproliferative substance (Wise, 1956). Alternatively, neovascularization could result from ischemic retina due to the loss of endogenous antiangiogenic factors either through their decreased synthesis, increased degradation, or inactivation. Further study of both developmental and adult ocular tissues will be necessary to determine the control points in ocular neovascularization.

164

ACKNOWLEDGEMENTS

The author wishes to thank Gerard Lutty for reviewing the
manuscript and for his major contributions to the migration and in
vivo studies. I am grateful to Dr. Patricia A. D'Amore for
sharing her knowledge of the retinal factor with me and for
providing samples of the factor for our tests. I also thank Dr.
Allan H. Fenselau for his support and help, especially with the
inhibitor reversibility experiments. Also, I wish to thank Dr.
Arnall Patz for his continued support and for assisting me in the
discussion of RLF. Portions of this work were supported by NIH
grants EYO2603 (A.P.) and CA 15381 (A.H.F.).

REFERENCES

Ashton, N., B. Ward, and G. Serpell. 1953. Role of oxygen in the
    genesis of retrolental fibroplasia: Preliminary report. Br.
    J. Ophthalmol. 37; 513-520.
Ashton, N., B. Ward, and G. Serpell. 1954. Effect of oxygen on
    developing retina vessels with particular reference to the
    problem of retrolental fibroplasia. Br. J. Ophthalmol. 38;
    397-430.
Auerbach, R. 1981. Angiogenesis-inducing factors: A review.
    Lymphokines, 4; Academic Press, 69-88.
Ausprunk, D.H., D.R. Knighton, and J. Folkman. 1975.
    Vascularization of normal and neoplastic tissues grafted to
    the chick chorioallantois. Am. J. Pathol. 79; 597-628.
Ausprunk, D.H. and J. Folkman. 1977. Migration and proliferation
    of endothelial cells in preformed and newly formed blood
    vessels during tumor angiogenesis. Microvascular Res. 14;
    53-65.
Brem, S., H. Brem, J. Folkman, D. Finkelstein, and A. Patz. 1976.
    Prolonged tumor dormancy by prevention of vascularization in
    the vitreous. Cancer Res. 36; 2807-2812.
Brem, S., I. Preis, R. Langer, H. Brem, J. Folkman, and A. Patz.
    1977. Inhibition of neovascularization by an extract derived
    from vitreous. Am. J. Ophthalmol. 84; 323-328.
Chen, C-H. and A. Patz. 1976. Components of vitreous-soluble
    proteins: Effect of hyperoxia and age, Investig. Ophthalmol.
    15; 228-232.
Clark, E.R. and E.L. Clark. 1932. Observations on living preformed
    blood vessels as seen in transparent chambers inserted in the
    rabbit's ear. Am. J. Anat. 49; 441-477.
Cliff, W.J. 1963. Observations on healing tissue: a combined light
    and electron microscopic study. Phil. Trans. R. Soc. Lond. B.
    246; 305-323.
D'Amore, P.A., B.M. Glaser, S.K. Brunson, an A.H. Fenselau. 1981.
    Angiogenic activity from bovine retina: Partial purification
    and characterization. Proc. Natl. Acad. Sci. USA. 78;
    3068-3072.
Duke-Elder, S. 1963. In "System of Ophthalmology III", part 1,
    C.V. Mosby Co., St. Louis. 141-153.
Fenselau, A. and R.J. Mello. 1976. Growth stimulation of
    cultured endothelial cells by tumor cell homogenates. Cancer
    Res. 36; 3269-3273.

Fenselau, A., S. Watt, and R.J. Mello. 1981. Tumor angiogenic
    factor: Purification from the Walker 256 rat tumor. J. Biol.
    Chem, 256; 9605-9611.
Flower, R.W., D.A. Blake, S.D. Wajer, P.G. Egner, D.S. McLeod and
    S.M. Pitts. 1981. Retrolental fibroplasia: Evidence for a
    role of the prostaglandin cascade in the pathogenesis of
    oxygen-induced retinopathy in the newborn beagle. Pediatr.
    Res. 15; 1293-1302.
Folkman, J. 1972. Angiogenesis in psoriasis: Therapeutic
    implications. J. Invest. Dermatol. 59; 40-43.
Folkman, J. and R.S. Cotran. 1976. Relation of vascular
    proliferation to tumor growth. Int. Rev. Exp. Pathol. 16;
    207-248.
Fournier, G.A., G.A. Lutty, S. Watt, A. Fenselau, and A. Patz.
    1981. A corneal micropocket assay for angiogenesis in the rat
    eye. Invest. Ophthalmol. Vis. Sci. 21; 351-354.
Gimbrone, M.A., R.S. Cotran, S.B. Leapman, and J. Folkman. 1974.
    Tumor growth and neovascularization: An experimental model
    using the rabbit cornea. J. Natl. Cancer Inst. 52; 413-428.
Glaser, B.M., P.A. D'Amore, R.G. Michels, A. Patz and A. Fenselau.
    1980a. Demonstration of vasoproliferative activity from
    mammalian retina. J. Cell Biol. 84; 298-304.
Glaser, B.M., P.A. D'Amore, H. Seppa, S. Seppa, and E. Schiffmann.
    1980b. Adult tissues contain chemoattractants from vascular
    endothelial cells. Nature 288; 483-484.
Goudie, R.B., J.C. Spence, and R.J. Scothorne. 1980. Do vascular
    clones determine developmental patterns? Lancet 1980-I.
    570-572.
Greenblatt, M. and P. Shubik. 1968. Tumor angiogenesis:
    Transfilter diffusion studies in the hamster by the
    transparent chamber technique. J. Natl. Cancer Inst. 41;
    111-124.
Henkind, P. 1978. Ocular neovascularization. The Krill Memorial
    Lecture. Am. J. Ophthalmol. 85; 287-301.
Jack, R.L. 1972. Regression of the hyaloid vascular system. An
    ultrastructural analysis. Am. J. Ophthalmol. 74; 261-272.
Langer, R. and J. Folkman. 1976. Polymers for the sustained
    release of proteins and other macromolecules. Nature (London)
    263; 797-800.
Lutty, G.A., B.M. Glaser, and A. Patz. 1981a. Inhibition of
    vascular endothelial cell chemotaxis by extracts of adult
    bovine vitreous. Invest. Ophthalmol. Vis. Sci. Suppl. 20;
    141.
Lutty, G.A., D.C. Thompson, J.Y. Gallop, R.J. Mello, A. Patz, and
    A. Fenselau. 1981b. Vitreous: An inhibitor of retinal
    extract-induced neovascularization. Submitted for
    publication.
McAuslan, B.R. and H. Hoffman. 1979. Endothelium stimulating
    factor from Walker carcinoma cells. Relation to tumor
    angiogenic factor. Exp. Cell Res. 119; 181-190.
Mello, R.J. 1977. Vascular endothelial cell culture: Use in
    studies of an endothelial cell mitogen derived from the Walker
    256 rat carcinoma. Doctoral dissertation. The Johns Hopkins
    University School of Medicine, Baltimore, MD.
Michaelson, I.C. 1948. The mode of development of the vascular
    system of the retina, with some observations on its
    significance for certain retinal diseases. Trans. Ophthalmol.
    Soc. UK. 68; 137-180.
Murray, P.D.F. 1932. The development in vitro of the blood of the
    early chick embryo. Proc. R. Soc. B. III; 497-521.
Muthukkaruppan, V.R. and R. Auerbach. 1979. Angiogenesis in the

mouse cornea. Science. 205; 1416-1418.

Patz, A. 1980. I. Studies on retinal neovascularization. Friedenwald Lecture. Invest. Ophthalmol. Vis Sci. 19; 1133-1138.

Patz, A., A. Eastham, D. Higgenbotham, and T. Kleh. 1953. Oxygen studies in retrolental fibroplasia: II. Production of the microscopic changes of retrolental fibroplasia in experimental animals. Am. J. Ophthalmol. 36; 1511-1522.

Sabin, F.R. 1917. Origin and development of the primitive vessels of the chick and pig. Carnegie Inst. Contrib. Embryol. 6; 61-124.

Sabin, F.R. 1920. Studies on the origin of blood vessels and of red corpuscles as seen on the living blastoderm of the chick during the second day of incubation. Carnegie Inst. Contrib. Embryol. No. 36. 9; 213-262.

Schoefl, G.L. 1963. Studies on inflammation. III. Growing capillaries: Their structure and permeability. Virchow Arch. Pathol Anat 337; 97-141.

Wagner, R.C. 1980. Endothelial cell embryology and growth. In "Adv. Microcirc". 9; (Karger, Basel) 45-75.

Weiss, J.B., R.A. Brown, S. Kumar, and P. Phillips. 1979. An angiogenic factor isolated from tumors: A potent low-molecular-weight compound. Br. J. Cancer 40; 493-496.

Wise, G.W. 1956. Retinal neovascularization. Trans. Am. Ophthalmol. Soc. 54; 729-826.

Yamagami, I. 1970. Electronmicroscopic study on the cornea. I. The mechanism of experimental new vessel formation. Jap. J. Ophthalmol. 14; 41-58.

Yanoff, M. 1969. Ocular pathology of diabetes mellitus. Am. J. Ophthalmol. 67; 21-38.

Theories of Magnification Relative to the Visually Impaired

Richard Brilliant and Sarah Appel

Low vision is a term that has been used in the field of eye care since the inception of the first eye care professional. It has also been called subnormal vision, partial sight and visual handicap. Regardless of the term used to describe vision loss, the realities of low vision confront a large number of individuals. Estimates range from 1.7 million individuals (Genensky, 1978) to 10,659,000 (Westat, 1976) people with serious disability due to visual impairments. The numbers are expected to climb yet higher as advances in medicine increase life expectancy, thereby increasing the incidence and severity of conditions such as diabetic retinopathy and senescent macular degeneration, major causes of reduced vision in the elderly.

Before proceeding with the theories behind low vision magnification systems it is important to define what degree of impairment constitutes a low vision individual. A distinction must initially be made between the terms low vision and "legal blindness". For purposes of state and federal tax exemptions, federal disability benefits, rehabilitation services, etc., legal blindness is defined to be 20/200 visual acuity or worse in the better eye with standard (non magnifying) correction, or a visual field no wider than 20 degrees.

Low vision, however, encompasses those with less severe vision loss as well as those who are legally blind. Unlike the definition of legal blindness, low vision cannot be defined by visual acuity or visual field boundaries. Some practitioners define low vision as those individuals who exhibit acuities of 20/70 or poorer through standard correction. This definition, however ignores vast numbers of individuals who possess better acuities but seriously impaired visual functioning. A realistic definition must consider the level of visual functioning with existing acuities. "It (low vision) may be defined clinically as the point at which the patient becomes aware that his poor acuity has affected his performance so that he thinks of himself as handicapped. It cannot be defined 'numerically'" (Faye, 1970). A good example would be that of an intelligent 26 year old woman who found it impossible to read even though she had 20/25 vision in her better seeing eye. Upon careful visual field testing it was found that although there were no limitations in peripheral fields, she had two small scotomas to the right and left of fixation that obscured the beginning and end of words. She certainly wasn't legally blind and yet she was unable to function in an academic setting. She was, therefore, considered a low vision patient, and benefited greatly from a thorough low vision evaluation.

The eye care professonal all too often encounters the frustration of dealing with ocular pathology that cannot be remedied by surgical intervention or medication. At best, in the case of nonprogressive conditions, the practitioner may assure the patient that total blindness will not occur. Few practitioners, however, are able to suggest techniques whereby visual functioning may be improved. With the development of the field of Low Vision Rehabilitation, practitioners employ optical systems combined with training techniques in order to maximize residual visual functioning. In a large percentage of cases this is accomplished through the use of magnification.

Without magnification an image may be distorted, fragmented, or totally obscured depending on the nature of the ocular defect as well as the size and location of the scotoma. For example, in the case of damaged retinal tissue, the involved photoreceptors are incapable of sending signals to the visual cortex. The number of photoreceptors that are functioning and transmitting stimuli may be inadequate to enable interpretation of the image. When the retinal image is enlarged through magnification, it encompasses a larger number of functional photoreceptors. The detrimental effects of nonfunctional photoreceptors contained within the boundaries of the retinal image are minimized and the stimuli relayed to the visual cortex are intensified. The object of regard is thereby rendered recognizable.

Magnification for the low vision patient is accomplished by increasing the angular size of a retinal image. Four methods are

employed to bring this about: I)   Angular Magnification;   II)
Relative Distance Magnification;   III)   Relative Size
Magnification;   and   IV)   Projection Magnification.

ANGULAR MAGNIFICATION

Angular magnification involves the use of a lens system to
increase the angular subtense of an image without physically
changing the position or size of that object.   It may be defined
as the ratio of the angular size of the image seen through the
lens system to the angular size of the object seen without the
lens system.

Low vision aids that create pure angular magnification are
afocal telescopic systems.   A simple telescope is comprised of 2
basic optical systems.   The front lens or objective is a
converging system.   Although a concave mirror may be used, the
common low vision telescope utilized a plus lens objective.   The
"eye piece" lens or ocular may be either a positive or negative
lens.   When the ocular lens and objective lens are separated at
the proper focal distance, parallel light rays entering the system
will also emerge parallel.   The lens system, however, will
increase the angle of emergence of the parallel light rays.
Magnification is the ratio of the angle of emergence to the angle
of incidence or

$$\frac{Tan^o e}{Tan^o i}.$$

Taking it one step further, another ratio that defines
magnification is

$$m = \frac{-D_{ocular}}{D_{objective}}$$

The magnification of a telescope is thus simply the negative ratio
of the dioptric power of the two lenses involved.

The telescopic system most commonly used consists of a
positive objective lens and a negative ocular lens (Figure 1).
This lens system is known as a galilean telescope.   Since the
ocular is a minus or concave lens (-D), m is a positive value.   A
positive m indicates an erect image.   Since galilean telescopes,
produce erect images, they are also known as terrestrial
telescopes.

In a galilean system, rays of light entering parallel through
the objective lens will come to focus at the posterior focal plane

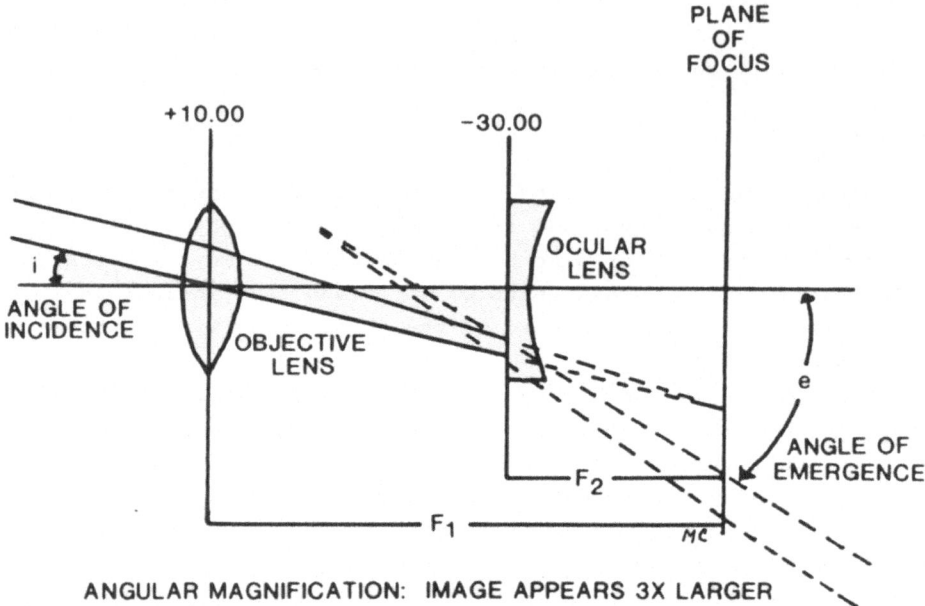

ANGULAR MAGNIFICATION: IMAGE APPEARS 3X LARGER

Figure 1. Diagram of angular magnification. See text for explanation.

of that lens. For light to leave the second lens in a parallel pencil of light, the rays must originate at the anterior principal focal plane of that lens. Therefore, to satisfy the condition of the two lenses, the anterior focal plane of the second lens must coincide with the posterior focal plane of the first. In describing the size of the telescopic systems (where the two focal planes coincide), there exists a relationship in which the sum of the focal lengths of the ocular and objective lenses (f_{oc} and f_{obj}) equals the tube length (d).

$$d = f_{oc} + f_{obj}$$

Since the ocular is negative in a galilean system the tube length is equal to the difference between the focal lengths.

$$d = (-f_{oc}) + f_{obj}$$

In a kepler telescopic system both lenses are positive. Since both D_{ocular} and $D_{objective}$ are positive values, m must be negative and the image is inverted. Tube length in this inverting or astronomical telescope is greater than that of a comparably powered galilean system since both lenses are positive and must be separated by a distance equal to their focal lengths.

Because of the increased tube length and the inverted image, kepler telescopes were thought to be useless as low vision systems. Erecting systems, such as a pair of porro prisms coupled with prisms that shorten the path of light rays within the telescope, have propelled the kepler system into prominence as a telescopic low vision aid.

The question may be raised of the need for the more complex kepler system when the terrestrial galiean already produces an erect image with a shorter tube length. The answer lies in a comparison of the field of view of both systems. A kepler telescope has a larger field of view than a comparably powered galilean system. This is due to the difference in the positioning of the systems' exit pupils. In a telescopic system the exit pupil is the image of the objective created by the ocular lens. All light rays passing through the objective lens must exit through the image of the objective lens or exit pupil. In a keplerian system the exit pupil is a real image and is situated in the space between the eye and ocular lens. Since the exit pupil and the eye are in close proximity much of the light entering the objective lens will enter the observer's pupil. The galilean system, however, produces a virtual exit pupil which is situated in the space between ocular and objective lens. The space between the exit pupil and the observer's eye is thus increased and the exit pupil reduces the amount of light entering the eye. The field of view is therefore smaller. Since field of view normally decreases as magnification increases, the kepler telescopic systems are the aids of choice in magnifications greater than 3x. Examples of telescopic systems are presented in Figures 2 and 3.

In discussing the practical aspect of magnification through a telescopic lens, when a telescope produces 3x magnification, a threefold increase in visual acuity is expected (i.e. 20/300 improves to 20/100). However, even with the increase in acuity, mobility while looking through the telescope is not possible. The limited field of view, magnification of motion, and exaggerated proximity of objects, renders safe navigation impossible. When a telescope is used, it must be employed as a spotting device only. In spotting activities, the patient must have enough vision and a sufficiently broad visual field to effectively localize the object and see its general outline without the telescope. Correct instruction in localization, scanning and focusing techniques increase the chances for successful use of these systems.

RELATIVE DISTANCE MAGNIFICATION

Relative distance magnification increases the angular subtend of an object by reducing the distance between that object and the observer. Magnification power may be described only in relation to the initial location of the object of regard (Figure 4). When an object, originally situated 40 cm from the eye, is moved to 10 cm, 4x magnification is achieved. If that object originated at 20 cm, only 2x magnification is achieved by moving it to 10 cm from the eye. It is well known that fixation distances closer than optic infinity (6 meters) require accommodation for clear vision, the shorter the distance, the greater the accommodative demand. If an accommodative response is not possible (i.e. aphakia, absolute presbyopia), or is not of sufficient magnitude (generally, fixation distances shorter than 10 cm) a plus (convex) lens must be provided to compensate for the lack of accommodative power. For example, in the case of a presbyope with 8 diopters of myopia or an emmetropic child who is able to accommodate 8 diopters, both can read at a fixation distance of 12.5 cm. Essentially both these individuals can call on 8 diopters of extra plus power to focus that image. An emmetropic presbyope, however, does not have access to any extra plus power in his visual system and would require a +8D spectacle Rx to see clearly at 12.5 cm. These plus lenses are referred to as low vision microscopes. Examples of these are presented in Figure 5.

Once the amount of magnification required for an individual to perform a nearpoint task is determined, the amount of magnification necessary may be related to the dioptric power of the required microscope by the formula $M_e = dF$ where M_e equals effective magnification, d equals reference distance (meters) and

Figure. 2. Telescopic Systems. Top: 2.8X Selsi Sport Glass. Middle Left: Selsi 6X Telescope. Middle Right: Walters 8X Short Focus Telescope. Bottom: 4X Selsi Binocular Sport Glass.
The above telescopic systems are readily available stock telescopes. The spectacle mounted system is advantageous for tasks involving continuous distance viewing (i.e. television, theatre, sporting events). Due to the central mounting of the telescopes, however, it is impossible to walk while continuously looking through it. Magnification will cause objects to appear to be closer and move with greater velocity, creating erroneous mobility cues. For general short term distance spotting tasks (i.e. street signs, but numbers, house numbers) the hand held systems are more functional.

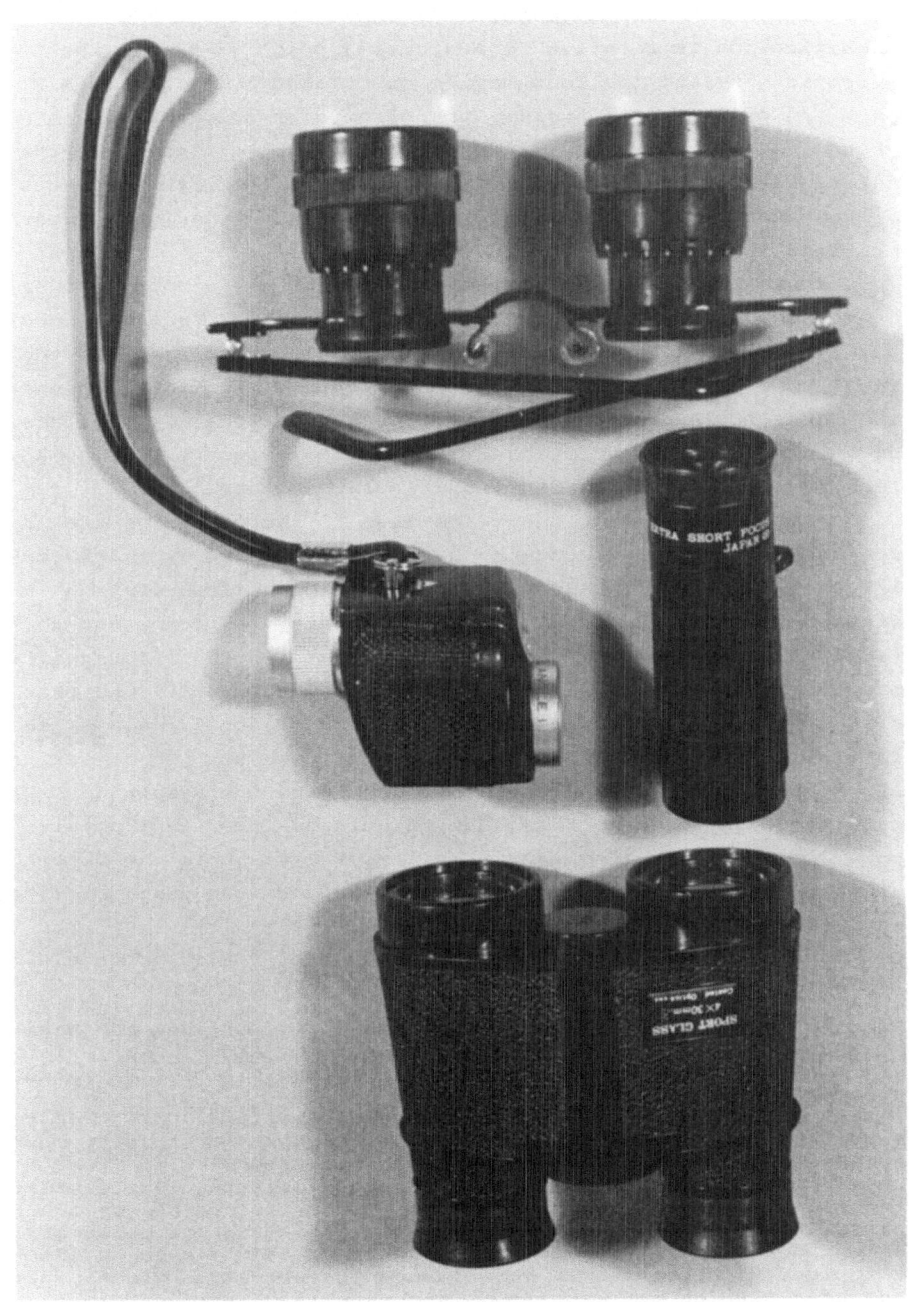

F equals the dioptric power of the microscope. For example, if an individiual requires 5x magnification to read and the magnification is relative to a starting point of 33 cm, the dioptric power of the lens may be calculated by the formula 5x = F/3 or F = 15 diopters. When the reference distance is not stated, a 25 cm distance is traditionally recognized to be the reference distance. Therefore, if the manufacturer designates a microscope or hand-held magnifier as a 4x magnifier, the power of the lens F equals +16.00D (M_e = F/4).

There is a direct relationship between the dioptric power of a microscopic lens and the distance for which it focuses (focal point). By using the formula f = 100/D, in which f equals the focal point of the lens (in cm) and D equals the dioptric power of the lens, it is possible to determine how far the material has to be held from the lens for the image to be in focus. For a +20.00 lens the correct focal point is calculated to be 5 cm.

Plastic, aspheric lenses are used to reduce the weight of the lenses while also reducing the distortion created by thick convex lenses. In order to reduce the problems of a close working distance, hand-held and stand magnifiers may be prescribed (Figures 6 and 7). The magnification that a hand-held magnifier provides is proportional to its dioptric power, as in the case of a spectacle microscope, assuming that the object is held at the focal point of the magnifying lens. If, however, the object is situated inside the focal point of the lens, the resulting magnification would be less than calculated from the dioptric power of the lens using the formula M_e = F/4. When the object is situated outside the focal point of the lens, the image will be inverted.

Figure 3. Bioptic Telescopic Systems. Top: Designs for Vision 3X Honeybee Bioptic. Middle: Designs for Vision 3X Keplerian Bioptic "Camera Lens". Bottom: Designs for Vision 2.2X Galilean Bioptic.

Spectacle mounted bioptic telescopic systems do not restrict general mobility activities to the degree that a centrally mounted spectacle telescopic systems would. A telescope is mounted in the upper portion of a conventional prescription lens. While walking, the individual views through the bottom portion of the standard prescription lens. When distance spotting tasks are necessary, a slight downward tilting of the head brings the telescopes into alignment with the visual axis. Once the object is discriminated, the head resumes its normal position and the individual once again views through the standard prescription carrier lens.

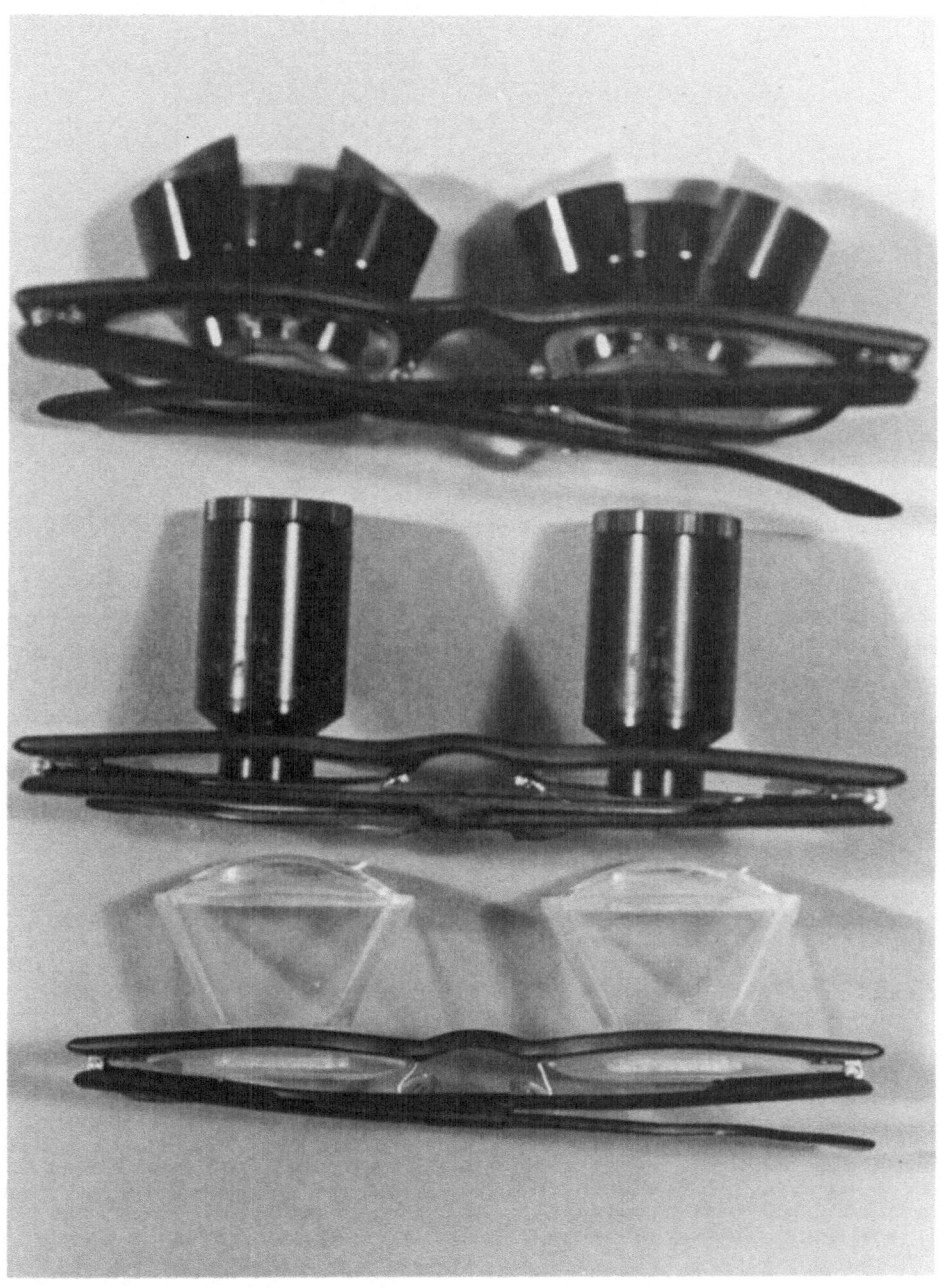

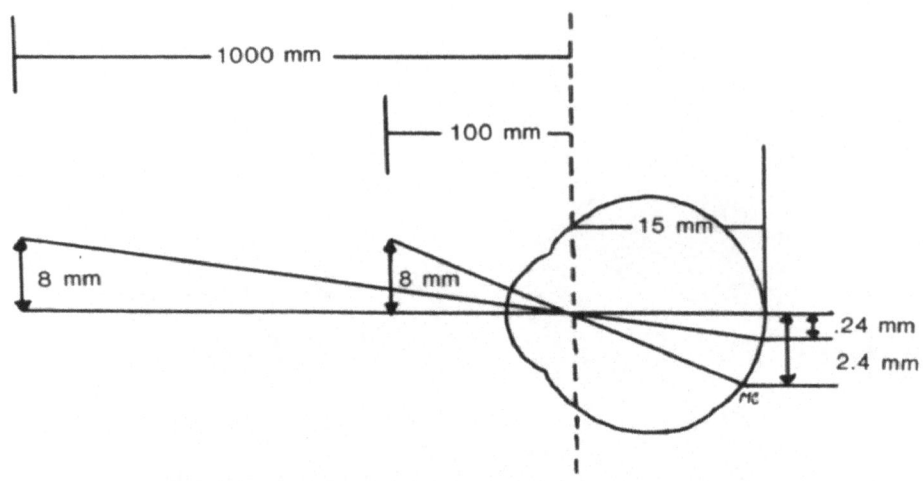

LINEAR MAGNIFICATION: IMAGE APPEARS 10X LARGER

Figure 4. Diagram of linear magnification. See text for details.

When a fixed stand magnifier is used there are two important points that must be considered. The patient must accommodate (or use an appropriate nearpoint correction) while looking through the system, and the magnification provided is always less than what the manufacturer has designated as the true power. This is due to the fact that the stand magnifier lens is set closer in than the focal distance of the lens. The light rays leaving this system are, therefore, not parallel, but divergent. The proper accommodative response or near correction will render them

Figure 5. Spectacle Microscopes. Top: American Optical 5X (20 Diopter) Plastic Lenticular Microscope. Middle: Designs for Vision 5X (20 Diopter) Microscope. Bottom: American Optical 8.00 Diopter Half-eye Microscopes with 10 Diopters of Base in Prism.
Spectacle mounted microscopes provide a larger field of view and free both hands for activities such as writing, drawing, and maintaining literature at the proper focal distance. Microscopes up to 3X magnification may allow binocular vision if base in prism is incorporated into the lenses to facilitate convergence for the closer working distance. Above 3X, binocularity is generally not achievable due to the tremendous convergence demanded by the short focal distances of the systems. A patch is generally worn over one lens as may be seen in the middle microscopic system. Disadvantages center mainly on the short focal distances of the higher magnification systems (5X and greater).

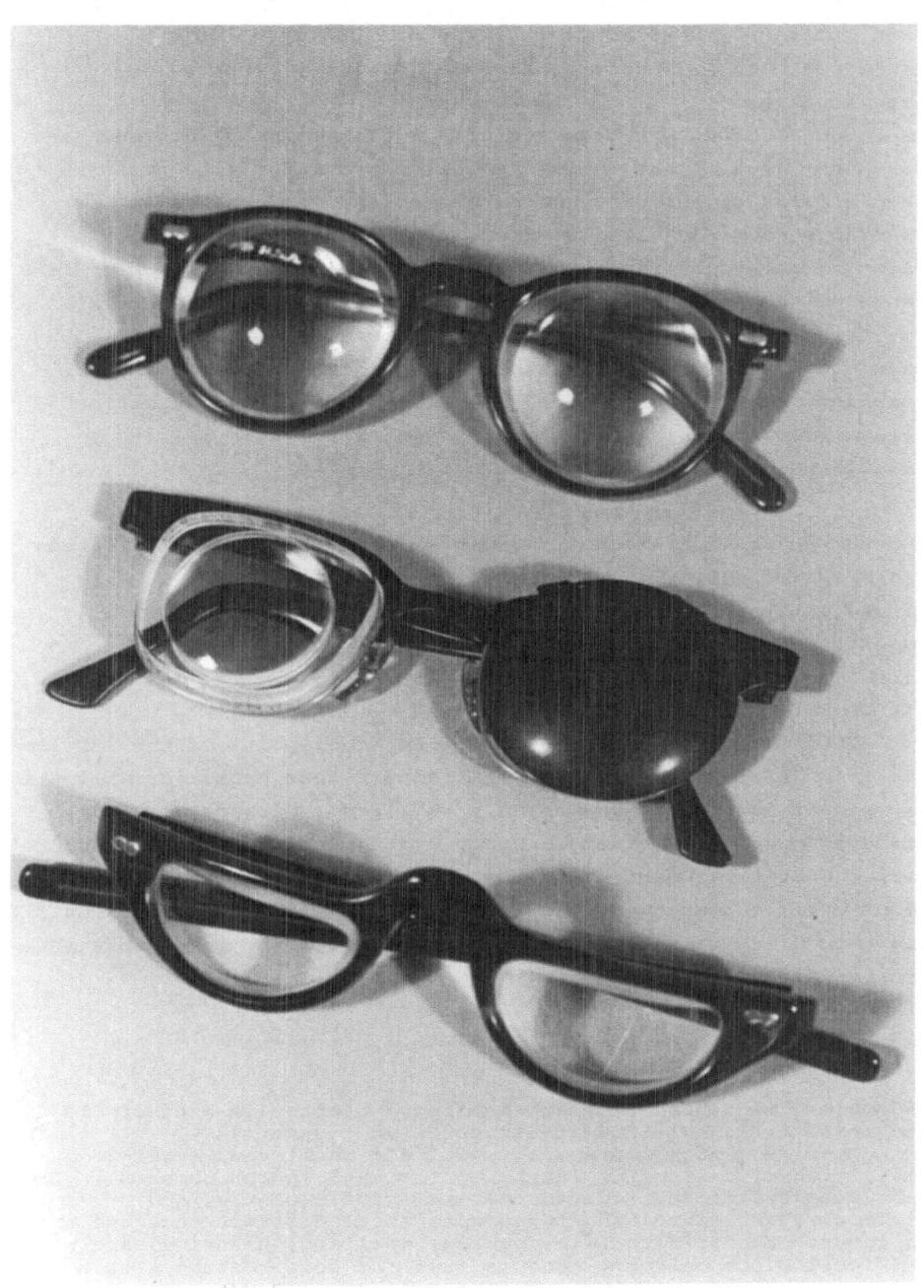

parallel, thereby clearing the image. As an example of actual
magnification versus manufacturer's designation of the
magnification, a 7x Coil stand magnifier only produces 5x
magnification.

When the patient requires both a greater focal distance and a
headborne system that would free his hands for nearpoint tasks
such as writing, playing an instrument and handicrafts, a
spectacle mounted telemicroscopic system may be indicated. A
telemicroscopic system is created by the placement of a plus lens
cap over the objective lens of an afocal telescope (Figure 8).
The telemicroscope will focus at the focal distance of the plus
lens cap. This system is essentially a combination of angular
magnification (telescope) and relative distance magnification
(plus lens cap). The magnification of this system may be
calculated by the following formula:

$$M_{tot} = (M_t) \times (M_e)$$

where M_{tot} is total magnification, M_t is telescope magnification,
and M_e is effective magnification of the lens cap (F/4).
If one combined a 4x telescope with a +5 ocular cap, total
magnification would be:

$$M_{tot} = (4) \times \frac{(5)}{4} = 5x$$

This 5x system would focus at the focal point of the +5D lens
cap (f = 100/D), or 20 cm. A +20D convex lens magnifier will also
provide 5x magnification but only at a focal distance of 5 cm.
Disadvantages of a telemicroscopic system include a small field of
view and critical depth of focus. Individuals, however, who
require large amounts of magnification at an increased working
distance often find telemicroscopic systems to be invaluable in
achieving their visual goals.

Figure 6. Hand Held Simple Magnifiers. Left: 3X Windsor (12
Diopters). Right: 5X Coil Aspheric (20 Diopters)
  Hand held magnifiers are useful for short term reading
activities (i.e. reading labels, price tags, phone numbers). Long
term use is difficult due to the small field of view and
difficulty in maintaining the proper focal distance once the arm
fatigues. Handicrafts (i.e. sewing, knitting, painting) are
difficult since only one hand is free and focal distance is short.

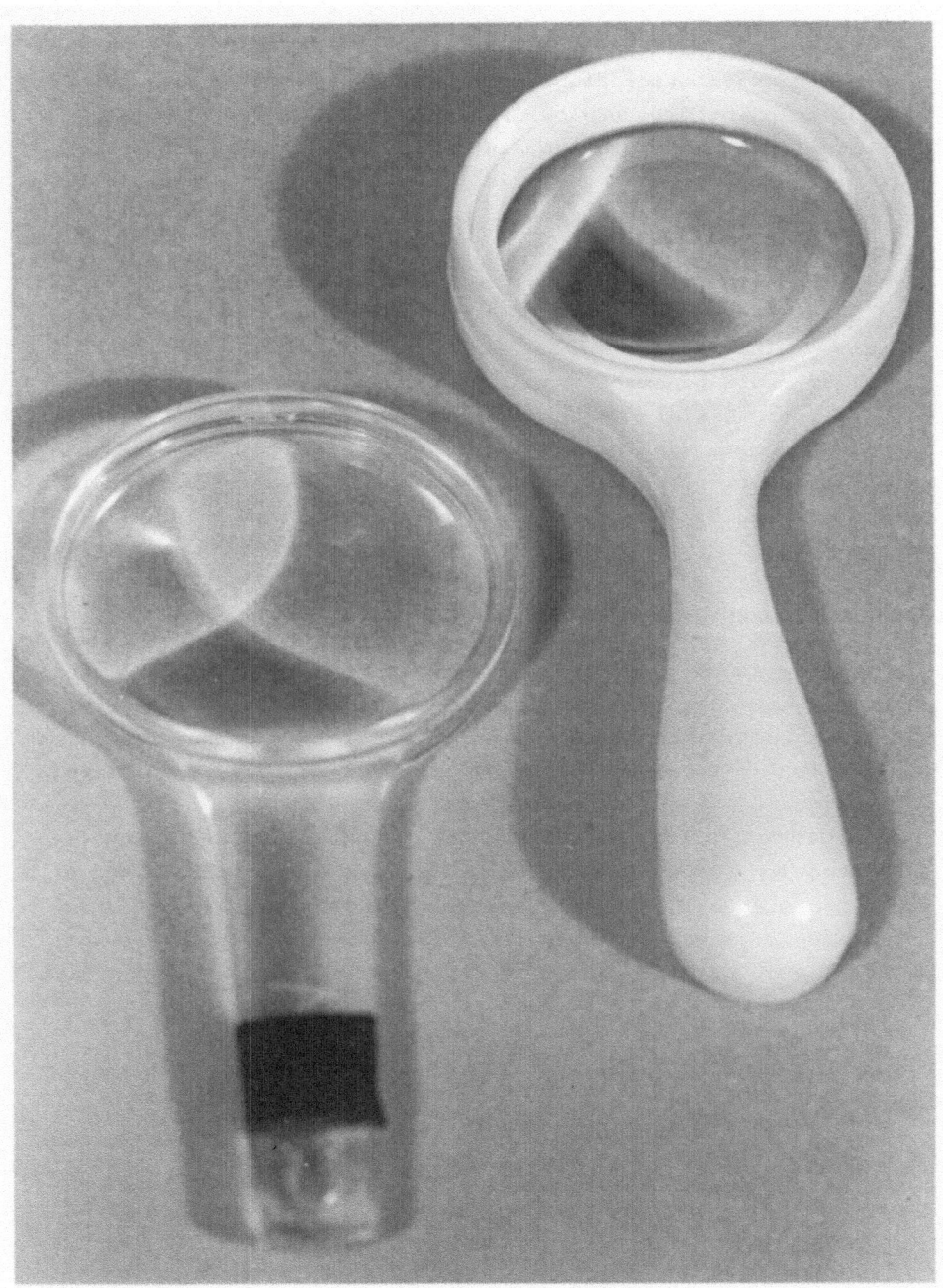

RELATIVE SIZE MAGNIFICATION

Relative size magnification involves increasing the physical size of an object (see figure 9). Increasing 20/20 size print to 20/60 size print will create 3x magnification. In low vision, such magnification may be obtained with a number of non-optical aids. The most common are large print books or newspapers, large phone dials, large-eyed sewing needles and large print playing cards.

Large print books or newspapers provide a wider field of printed matter while at the same time allowing a greater reading distance. Individuals who have not read for many years find it useful to begin practicing on large print materials with or without optical aids. Discrimination of words is easier as there tends to be more spacing between letters, words and lines when using large print. Large print materials, however is bulkier than standard print size materials and is often not as available to the general public.

PROJECTION MAGNIFICATION

Projection magnification involves projection of the image of an object onto a screen. This type of magnification is achieved through a combination of electronics and lenses rather than with lenses alone. Excellent examples are seen in the closed circuit television (CCTV) and the slide projector.

A CCTV is indicated for a patient requiring large amounts of magnification (generally greater than 10x) and wanting a normal working distance for reading and writing. It may also be indicated for those individuals who require an overall view of material such as maps and photographs.

Figure 7. Stand magnifiers. Left: 9.00 Diopter Jupitor Stand Magnifier. Top Right: 20.00 Diopter Coil Stand Magnifier. Bottom Right: Agfa 32.00 Diopter Stand Magnifier.
Stand magnifiers reduce the fatigue factor since their focal distance is fixed. The magnifier simply slides across the reading material. Disadvantages, however, include a small field of view, short focal distance and the necessity for manual manipulation.

Advantages of this type of magnification are:

1) very large amounts of magnification can be obtained

2) the viewing or reading distance is at a normal reading distance

3) brightness and contrast of the image may be controlled

4) background illumination may also be controlled (i.e. black on white or white on black)

5) binocularity is possible with high amounts of magnification

6) there is a reduction of aberrations and distortions as compared to an equivalent powered microscope

7) adaptations can be made for typing

Disadvantages of the CCTV are:

1) it is not portable

2) maintenance may be a problem

VISUAL FIELDS IN RELATION TO MAGNIFICATION

In considering the prescribing of a magnification system for the low vision individual, the practitioner must be aware of visual field as well as visual acuity data. The patient may be placed in one of three categories:

1) vision reduction without visual field involvement

2) vision reduction with central field involvement

3) vision reduction with severe peripheral field loss

The first category is typified by conditions that involve hypoplasia of the macular region or cloudiness of ocular tissue and media. Albinism, congenital cataracts, and congenital nystagmus all result in lack of development of the macular region thereby causing acuity loss without accompanying central scotomas.

Figure 8. Bioptic Telemicroscopic Systems. Top: Designs for Vision 3X Keplerian Bioptic with Reading Cap. Bottom: Designs for Vision 2.2X Galilean Bioptic with Reading Cap.

Telemicroscopic bioptic systems provide a considerably longer focal distance than a comparably powered microscope. Intermediate distance activities (i.e. card playing, painting, reading sheet music while playing a musical instrument) are greatly facilitated. Disadvantages include a small field of view and more critical working distance.

184

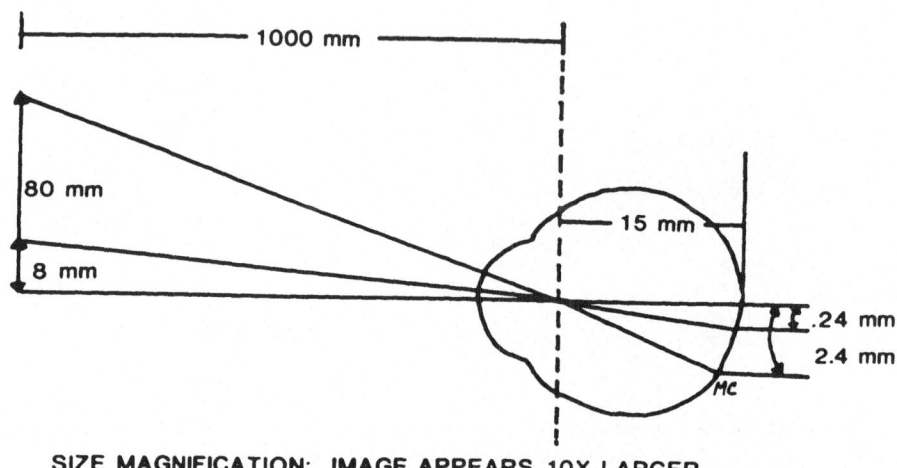

SIZE MAGNIFICATION: IMAGE APPEARS 10X LARGER

Figure 9. Diagram of size magnification. See text for explanation.

These individuals often respond best to magnification systems as there is no scotoma that may interfere with resolution of the magnified image. Pathologies affecting the transparency of ocular tissue and media include some forms of adult onset cataracts, vitreal opacaties and corneal dystrophies. Successful use of magnification in these pathological entities is contingent upon severity of involvement. When opacification is severe, magnification systems are ineffective and surgery is the only alternative.

Category #2 involves actual damage to photoreceptors or optic nerve fibers that receive stimuli from areas in the central 30° region of the visual field. The defect produces a scotoma in the visual field area. These pathologies include juvenile and adult onset macular degenerations, macular holes, toxoplasmosis, histoplasmosis, inverse retinitis pigmentosa and diabetic macular involvement. These individuals should also respond to

Figure 10. Apollo Closed Circuit Television Magnification System.

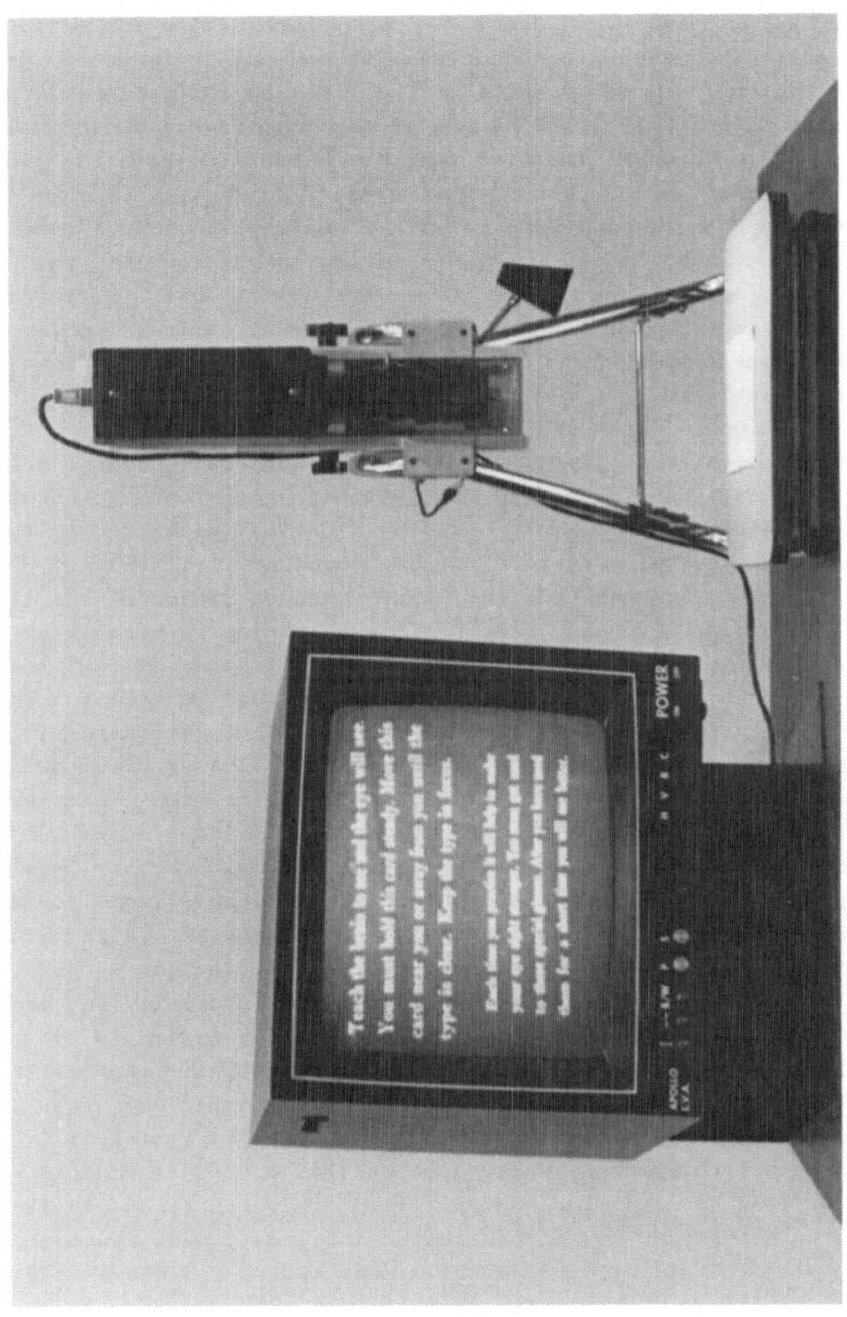

magnification but generally require more intensive training in
eccentric viewing techniques. In order to minimize the effect of
the scotoma on visual functioning, the patient is taught to use a
non-involved paramacular area as the fixating retinal region,
thereby effectively shifting the scotoma out of the direct visual
axis. Large central scotomas that are associated with the visual
acuity poorer than 20/400 (20° or greater) generally tend to
reduce the chances for success with magnification. As a result of
greatly increased magnification needs (greater than 10x) the
available optical systems are more difficult to use. Microscopes
have extremely small focal distances, thereby causing problems
with proper illumination as well as comfort. Telescopes have
significantly smaller fields of view that complicate proper
localization and scanning techniques. The increase in
magnification creates an increase in the apparent speed of the
retinal image causing marked problems with smearing of the retinal
image. A common complaint voiced by a patient experiencing this
phenomenon is "the words are running together". If that
individual is trained to significantly reduce motion of the image,
then a problem occurs in that peripheral retina is more likely to
exhibit a fading of the image due to stabilization of that image.
One cannot overstate the importance of thorough instruction in the
use of visual skills to solve the problems that patients
encounter. With appropriate training and follow-up some low
vision individuals have adapted successfully to magnifications as
high as 20x.

The final category includes individuals with acuity loss
coupled with severe peripheral visual field loss. Individuals who
have retinitis pigmentosa with less than 5° central island of
vision remaining, end-stage glaucoma, central retinal artery
occlusion or massive retinal detachment will generally not benefit
from magnification. Since the majority of the retina is
nonfunctioning, increasing the size of the retinal image will
magnify it onto atrophic tissue, and thereby interfere with
resolution of the whole image. In some cases of retinitis
pigmentosa and glaucoma where good central acuity is coupled with
10° or less of remaining central field area, minifying devices or
reverse telescopes may be considered instead. These systems
enable an individual with severely constricted peripheral fields
to see a larger visual field area.

In determining the appropriate low vision aid or combination of aids for a patient, it is essential that the practitioner consider the pathology, visual field and visual acuity data. Low vision care, however, is not based solely on the ocular status of the patient. The practitioner must also be aware of the physical, intellectual and emotional condition of the patient. Systemic complications such as arthritis or Parkinsonism interfere with proper manipulation of the hand-held magnifiers. They may create difficulties in the use of microscopic and telemicroscopic spectacles due to the increased problem in maintaining the proper focal distance. Individuals with severe memory problems or marked mental retardation may find the majority of low vision systems to be too difficult to master.

Since low vision systems tend to be task specific, the practitioner must be sensitive to the visual objectives of his patient. Unless an aid is prescribed in accordance with the patient's specifically stated needs, and unless the patient's interest and motivation are sufficient, mere clinical success does not constitute overall success. A practitioner who dispenses a microscopic system for reading to an individual who only wants to view television programs has failed to improve the quality of life for that individual.

With a basic understanding of the low vision patient, the four methods of magnification and the magnification systems that are available, eye care practitioners should be able to enhance the visual efficiency of many low vision patients. As the life expectancy of the average individual increases, there will consequently be increasing numbers of low vision patients. There are obviously many options available to help maintain a person's independence and hopefully patients will not hear the usual comment, "Sorry, nothing else can be done".

REFERENCES

Faye, E. 1970. The Low Vision Patient, New York, Grune & Stratton.
Faye, E. 1976. Clinical Low Vision, Boston, Little, Brown & Co.
Fonda, G. 1981. Management of Low Vision, New York, Thieme-Stratton, Inc.
Genensky, S.M. 1978. Data Concerning the Partially Sighted and Functionally Blind. J. Vis. Impair. Blind 5.
Mehr, E., and A. Freid. 1975. Low Vision Care, Chicago, The Professional Press
Kelleher, D. 1979. Orientation to Low Vision Aids, J. Vis. Impair. Blind. 73:161-166.
Sloan, L. and S. Ryan. 1971. Recommended Aids for the Partially

188

Sighted, National Society for the Prevention of Blindness.

Sloan, L. 1977. Reading Aids for the Partially Sighted, Baltimore, Williams and Wilkins Co.

Westat, Inc. 1976. Summary and Critique of Available Data on the Prevalence and Economic and Social Costs of Visual Disorders and Disabilities. Rockville, Maryland: Westat, Inc. USHEW, Public Health Service, Contract #1 EY 52108.

Index